GERMAN SCIENCE

PIERRE DUHEM

German Science

Some Reflections on German Science
German Science and German Virtues

Translated from the French by John Lyon
Introduction by Stanley L. Jaki

OPEN COURT
Chicago and La Salle, Illinois

To order books from Open Court, call toll-free 1-800-815-2280 or visit our website at www.opencourtbooks.com.

This book has been reproduced in a print-on-demand format from the 1991 Open Court printing.

Open Court Publishing Company is a division of Carus Publishing Company.

Printed and bound in the United States of America.

Library of Congress Cataloging-in-Publication Data

Duhem, Pierre Maurice Marie, 1861–1916.
  [Science allemande. English]
  German science : some reflections on German science : German science and German virtues / Pierre Duhem ; translated from the French by John Lyon ; introduction by Stanley L. Jaki.
      p.   cm.
  Translation of: Science allemande.
  Includes bibliographical references and index.
  ISBN 0-8126-9123-7 (cloth). — ISBN 0-8126-9124-5 (pbk).
  1. Science—Germany—History. 2. Science—Philosophy—Germany—History. I. Title.
Q127.G3D8513 1991
509.43—dc20   91-21454
                                                        CIP

CONTENTS

TRANSLATOR'S PREFACE

I fear I have nothing of significance to add to the interpretation of either *German Science* or the life's work of Pierre Duhem. Those tasks I gladly leave to others more knowledgeable than I. Perhaps a few words of caution to the reader would be in place, however.

The general reader is rare, the gentle reader scarcely exists today. Those who closely peruse these pages will tend to be academic specialists, professionally beyond ingenuousness, scandalized only by an apparently quirky passage here and there that crosses the threshold of their subconscious predilections. But the rare general reader may be brought up short by some of Duhem's prose. For instance: when the author concludes the fourth lecture by identifying 'Frankness' in speech with the injunction to Christians to let their yeas be yea, their nos, no; when he moves from a tribute to the balance and perfection in science of the Germanic Clausius, Helmholtz, or Karl Friedrich Gauss (Lecture IV) through a hasty reduction of "German science" to the generalization, "*scientia germanica ancilla scientiae gallicae*" (conclusion, 'Some Reflections on German Science'); or when, in Lecture IV he teases the Germans with catlike malice about moving logically in their patriotism from desired conclusion to requisite premises, or about their schizophrenic love of idealism on the one hand, and the solace they find in pipe, kraut, and beer on the other—when he does these things, we wince in recognition of the truth and in pain at the falsity, and reflect: 'Oh yes. Spring, 1915. Propaganda. Patriotic gore.' And then, perhaps unconsciously, we tend to discount all else Duhem has to say, for we latter-day sinners find it hard to forgive those Christians—French or German—whose patriotism was so severe that it led them to defend to the death—with all the ambiguity of which that phrase is endowed—all that remained of Christendom, and much of what remained of Western civilization.

We tend to discount Duhem's conclusions despite his
obvious attempt at fairness, despite the mildness of his critique
of things Teutonic compared to other contemporary vituperative-
ness, despite, to vary Stanley L. Jaki's words in the 'Introduction', the absence of extreme chauvinism. But there is obviously
much of value in this work that ought under no circumstances
to be discounted. We treasure, for example, Duhem's telling
characterizations delineated in passages such as those in pages
92–93 (*"wir können und wollen setzen"* . . . *"sic volo, sic jubeo,
sit pro ratione voluntas"*) in 'Some Reflections' concerning the
irrational, volitional aberrations of the exclusively deductive
mind. (Today, such half-random, half-disingenuous positing of
premisses in axiomatic systems we tend with some appropriate-
ness to call 'game-playing'.) We cherish his good sense about
history ("There is not, there cannot be, any historical method,"
Lecture III). We particularly appreciate the telling observations
at the very end of Lecture III concerning the consequences of a
powerful deductive intellect deprived of common sense, holding
as true "every consequence deduced by rule" from arbitrarily
postulated principles, principles—unconsciously derived perhaps—
designed to serve a cause. It is indeed as Duhem notes:
"Among scoundrels, this is most dangerous. Against remorse
they have laid hold of an assurance as certain as that two and
two make four."

Yet we still may tend to discount Duhem's effort, and not
only because of its patriotism which, to our sensitivities, is
exaggerated. We may do so, for instance, because we see him
involved in some apparent inconsistencies. We see him warning
his French audience against the Teutonic universities' ultimate
appeal—*Magister ipse dixit*—while he himself issues his own
'ipse dixits'. Again, we see him speaking out against that
supposed aberration of the Teutonic mind from Nicholas of
Cusa to Hegel (see the end of Lecture I) that starts from the
"sophistic" premise of the *coincidentia oppositorum*, while
his 'national' hero, Pascal, did not hesitate to propose that

"the sciences have two extremes which meet," nor to make further suggestions *de docta ignorantia* (cf. No. 327, Trotter translation, *Pensées*).

The specialized reader will pick and choose his way through this volume, and I wish him well in his search for the arcanum of his trade. But the rare general reader—if he is in any way like this odd, general translator—will set aside the obviously caricatured straw-men that Duhem sets up and blows down, and concentrate on the remarkable fairness that the author is capable of when, for example, he deals (Lecture IV) with true scientific genius—a *genius* that knows neither geographic nor national *loci*, but whose only place is in the human mind. And above all, I hope that such a reader will find Duhem's meditations on Pascal's distinctions between *l'esprit géométrique* and *l'esprit de finesse*—meditations that provide the theme and thesis that hold the miscellaneous contents of this small volume together—alone worth the price of admission.

We too might turn in hope to the spirit of Pascal who, when writing of Descartes and Montaigne, the great enemies he strove to overcome, could be catholic enough to recognize himself in them. "No one calls another a Cartesian but he who is one himself," he noted: or again: "It is not in Montaigne, but in myself, that I find all that I see in him" [*Pensées* (Trotter translation). New York: Dutton: Nos. 52, 64]. Though, like the author of the *Pensées*, I must confess to a dislike for "a mathematician" (for he will tend to mistake me for a proposition), I find that mathematician inextricably bound up in myself. The Rhine that separates us—from each other, as within our selves— is not the river that cascades at Schaffhausen. The Gaul that provides the true homeland for science is not the Gaul Caesar found divided into three parts. *Nous sommes tous des Boches.* One trusts Duhem knew that.

ACKNOWLEDGMENTS

It is my privilege to thank those without whose co-operation this translation would not have been possible. Stanley L. Jaki suggested over a decade ago that the work merited attention and proposed its translation to me. Over the intervening years, and across a most treacherous terrain in my own life, his constant pastoral care shepherded the work toward publication. But I am indebted to him for far more than his solicitude on behalf of this slender translation. The body of Jaki's work, and indeed the graceful asceticism of his life, have been a powerful inspiration to me (as they have been to so many others), and I am grateful for the opportunity of paying this tribute to him.

My work was aided by a grant from the Marguerite Eyer Wilbur Foundation, for whose assistance on this and on other occasions I remain sincerely grateful. To Dr. Russell Kirk, President of the Foundation, and to his wife and constant collaborator Annette Kirk, I owe more than words in a preface can decorously express. The achievement of each of their lives is extraordinary, their examples inspiring, their charity deep and orthopractic. Though the distinguishing features of their personal attainments are theoretically discriminable, they are in fact conjugally inseparable, and their 'corporate worth' is monumental. I have long stood in deep debt to their spiritual and mental credit, their social and civic magnanimity, and their personal grace.

Many others have had a hand in the prenatal care of this infant work, including David Ramsay Steele of Open Court, who has solicitously and sympathetically co-ordinated the professional details of publication for *German Science*. Dr. Bernard Doering of the Modern Language Department at the University of Notre Dame has regularly assisted me in an attempt to move my journeyman translation skills toward mastership—and has done so in a spirit of personal friendship which I treasure.

Ms. Anne S. Williams, of Seton Hall University, transformed my much-marred typed copy of the translation into a quite clear word-processor hard copy, for which I owe her great thanks. But chief among these others stands Niall Martin of Edinburgh, independent scholar and skilled interpreter of Duhem's work. Dr. Martin closely read my entire translation and kindly commented on troublesome passages, while regularly offering valuable stylistic advice. I feel that there were occasions on which a skill equivalent to that of rightly interpreting the flight of crook-taloned birds was requisite to elicit the relationship between passages in my proposed translation and congruent segments of Duhem's French. I am most appreciative of Martin's ability, his concern, and his skillful augury, all of which he has generously put at my disposition.

I must make payment of a final debt here, though I can never fully discharge it. Time and energy that were owed to my children have been regularly stolen from them over the years and applied instead to the completion of this and other similarly frivolous ventures. Age cannot repay what youth has taken away, and sins of inappropriately sequencing the stages of one's life, though forgivable in an economy of grace, yet remain indelible mistakes in the economy of nature.

To my children, then, in partial payment of that of theirs which I misappropriated, I dedicate this book: to each, by name, I send this symbol of reimbursement: of Thomas, Siobhan, Nora, Matthew, Geoffrey, Mark, Arthur, Kathleen, Sean, Jean, Christopher, Brendan—and of the one who lived not long enough to bear a name—I ask forgiveness.

John Lyon
Hillsdale College
June 1991

INTRODUCTION

Stanley L. Jaki

In the summer of 1916 Pierre Duhem jotted down the brief sketch of a book which he did not live to write. On September 16th of that year a massive heart attack[1] put an end to one of the most creative lives in modern times. Overwork was one of the reasons for a relatively early death at fifty-five. Between 1884, when in his second year as a first-ranking student at the École Normale Supérieure Duhem began to publish as a physicist, and his death 32 years later, his publications increased to well over twenty thousand printed pages. They all evidenced incisive learning not only in theoretical physics, his chosen field, but also in the philosophy and history of physics. In his time Duhem was a leading thermodynamicist with such lasting achievements to his credit as the Gibbs-Duhem equation. His researches in the physics of fluids have recently attracted the interest of students of plasma physics. As a philosopher of science he has remained an authority through his *Théorie physique* whose richness and originality is still to be fathomed in full. His single-handed discovery and massive documentation of the medieval origins of classical mechanics made him the creator of the modern historiography of science which aims at something far better than the clichés bequeathed by Bacon, Condorcet, Comte, and their latter-day allies.

In addition, Duhem was a most conscientious teacher, always at the disposal of his students. He first taught in Lille (1887–93), then in Rennes (1893–94), and finally in Bordeaux (1894–1916). That in spite of his renown he had been denied a chair in Paris was another cross for Duhem to bear, though not the heaviest of his crosses. The few glimpses he had revealed of his sense of bereavement (his wife died in childbirth in their third year of marriage in 1892) may help form an idea of the burden with which his heart could not cope in the long run. But in 1916 Duhem's major heartache had nothing to do either with academic or with family matters. By then he had for eight years coped with living as a solitary. His daughter and only child, Hélène, had, since 1908, been away from home for the better part of each year in Paris where, as an auxiliary to a charitable organization, she helped young working women in the banlieux. For a patriot like Duhem, frustrated by his inability to see active service and to send a son to the battlefront, the summer of 1916 was a particularly depressing time. The Germans had just been stopped at Verdun, but at the cost of half a million French casualties. Almost as many were to be the price of the unsuccessful French counter-offensive on the Marne which began in the early summer of 1916. Not only France but Europe was bleeding white.

Even those far from the front could not help sensing that World War I was something new in the history of warfare. Until then man fought man and therefore the carnage was limited. But in World War I carnage for the first time became mass-produced through heavy reliance on new forms of weaponry such as gas, machine guns, tanks, and submarines. War was no longer so much a question of personal bravery as a matter of technological excellence which largely depended on the harnessing of science in the cause of warfare. It was an open secret that chemical laboratories on both sides played a crucial role in the prolongation of war. Synthetic nitrate, a chief ingredient of gunpowder, had to be invented both in Britain and in Germany, because natural nitrate, mostly mined in Chile, became inaccessible to both sides.

The colossal carnage science helped to produce was in Duhem's eyes an abuse of science amounting to the gravest sin which for Duhem, the devout Catholic, was the sin against the Holy Spirit, a sin barring forgiveness. Such was the gist of the book Duhem planned to write in the summer of 1916. It would have been a sequel to the book, here translated, and like this book, it would also have been delivered in the form of public lectures under the auspices of the Association Catholique des Étudiants de l'Université de Bordeaux.

It would, of course, be wholly mistaken to think that Duhem would have become a pacifist advocating a unilateral laying down of French arms. He was no daydreamer forgetful of the real world. As a Christian he knew that the real source of the problem was not science and technology, but man's fallen nature. He also knew that although this dogma of original sin was the experimentally most evident of all Christian dogmas (to paraphrase a dictum of Chesterton), it was also the dogma most resisted by the world. Little impact was in fact made when in the wake of Hiroshima the momentous warning came from the non-Christian side (Einstein to be specific) that not the uranium but man's heart needed purifying in the first place.

So much for a theme to which this book might have become a worthy introduction. For, as this book shows, Duhem was as much a master of thought as he was a master of the pen. Twenty years before this book was written Duhem was on his way to becoming the chief figure of the *haute popularisation* of science in France. But following the success of the first three articles in a series which he was invited to write on the new science of thermodynamics for the *Revue des Deux Mondes*, the most prestigious biweekly in France, if not in the entire world, the editor of the *Revue* was pressured from 'higher places' to inform Duhem that the series was to be terminated in mid-course. The 'higher places' meant the power circles ruled by

Marcelin Berthelot, who added to his prominence as experimental chemist the role of a pontiff in the Third Republic.

Berthelot's resentment of Duhem (34 years his junior) and the main thrust of this book, a spirited defense of French learning and mentality, are not at all unconnected. For it was out of a sense of patriotic duty that young Duhem chose in 1884 for his doctoral dissertation a topic which not even well-established French scientists dared to discuss in public at that time, and even later, for fear of Berthelot's power that could be asserted in any academic decision relating to the French university system and well beyond it. Berthelot, the experimental chemist, did not refrain from serving in two cabinets, first as minister of the interior and then as minister of foreign affairs. His hold over the Académie des Sciences and the *grandes écoles* in Paris was a byword. His chief scientific pride was the so-called maximum work principle, which he was suspected of having borrowed from Thomsen, a Danish chemist. Although a good practical rule, the maximum work principle lacked a sound theoretical basis. Yet Berthelot succeeded in making that principle widely accepted in France from the 1870s on.

It is that influence, so harmful to the progress of physical chemistry in France, that young Duhem was resolved to undermine with his doctoral dissertation. Its subject was the thermodynamic potential, his own brainchild, which implied a refutation of the maximum work principle. The dissertation, which contained the formula later called the Gibbs-Duhem equation and today is part of the series *Landmarks of Science*, was rejected by the Sorbonne. Young Duhem was also made to understand that he was never 'to arrive in Paris'. He spent in fact all his academic career in provincial French universities, an obvious exile for one of Duhem's excellence. No slight, however, could break Duhem's spirit and resolve. Less than ten years later, while still in Lille, Duhem began publishing his views on the cultivation of science, or physics to be specific, in a genuinely French spirit. For Duhem this meant both a compliance with the dictates of rigorous demonstration, in which every mathematical step had to correspond to some physical reality, and a mistrust of imagination bent on model making. The latter was, in Duhem's eyes, the distinctive feature of the Anglo-Saxon mind. Fertile as imagination could be with respect to making discoveries, Duhem was ready to leave that glory to the Anglo-Saxon provided he could claim for the French the glory of turning the wealth of discoveries into a secure system.

Tellingly, Duhem saw at that time and even a dozen or so years later when he published his *Théorie physique*—his mature reflections on the aim and structure of physical theories—a great affinity between the French mind and the German mind, precisely because both were mainly interested in erecting rigorous systems. Such a sympathetic view of the German mind was common in France in the decades preceding World War I. It grew in part out of French reflections on France's defeat in 1870 and on the spectacular rise of science in Bismark's Germany. In fact, as late as May 1914, no less a representative of French learning than E. Boutroux praised the unity of German and French minds in a widely publicized lecture delivered at the University of Berlin. In May 1914 Boutroux's was a conspicuous case of myopia, the origin of which went back to the panegyrics which Renan, another chief figure of the Third Republic, sang half a century earlier about German culture.

Fortunately for Duhem he had for director at the École Normale no less a historian than Fustel de Coulanges, for whom he kept a lifelong respect (witnessed by the third lecture in this book). The pleas of Fustel de Coulanges for a balanced view concerning German achievements found a receptive soil in Duhem. Never a worshipper of Germany, he was not to become filled with hatred of it even in the darkest hours. This demanded courage on the part of a Frenchman in late 1914 when Duhem accepted the invitation of the Abbé Bergereau on behalf of the Association des Étudiants Catholiques of Bordeaux University to deliver four lectures on German science on four consecutive

Thursdays between February 25 and March 18, 1915. The
Association, established in 1913, had Duhem as one of its
founding members. Duhem readily lent to the Association his
prestige which greatly increased when in late 1913 Paris
'capitulated' to him through his election as one of the first six
non-resident members of the Académie des Sciences. In turn,
Duhem found much relief from his solitary status in this new
and informal contact with students for whom he quickly
became not so much a revered father-figure as a *camarade* ready
to relive his youth among the young.

The lectures from the very start drew an overflow audience
that could not be accommodated by the main hall of the head-
quarters of the Association. Owing to the rigid separation of
Church and State the lectures could not be transferred to a
large auditorium at the University. The last three lectures were
delivered in a theater nearby. The lectures, widely reported in
the Bordeaux newspaper, were quickly printed as a volume
which included an article Duhem was asked to write on German
science shortly beforehand for the *Revue des Deux Mondes*. Gone
were the days of the establishment's ostracism of Duhem.

The present translation also includes an essay on the subject
which Duhem contributed a few months later to a collective
work in which two dozen leading French intellectuals discussed
the respective merits of German and French culture. Such a
discussion was considered a patriotic duty, a contribution to
strengthening national resolve. World War I was a first in the
history of warfare not only because of carnage raised to gigantic
proportions, but also because of the role given to ideology.
Scientists had to take part in the propaganda warfare no less
than writers, artists, and politicians. The days were long past
when a Humphrey Davy, best remembered as Faraday's
mentor, commented on his free travel through France, at a time
when France interned British citizens and Britain interned
Frenchmen, with the words: ''If two countries are at war, the
men of science are not. That would be a civil war of the worst

possible description." A hundred years later scientists were engaged in a war "of the worst possible description" not only by offering their technical know-how but also their ideological interpretations of it.

As one could expect, a good deal of the ideological literature flourishing on both sides was full of venom and vituperation. In the eyes of Germans all Frenchmen suddenly became so many Jacobins and in French eyes all Germans so many Teuton savages. This rather 'below the belt' tactic made itself apparent on occasion even in publications where effort was made at maintaining civility. Thus, in the book to which 'German Science and German Virtues' was first contributed, a French medical man claimed, in reminiscing on his visit to German children's hospitals and orphanages, that German babies were invariably manhandled by their nursemaids.

Duhem was not to allow himself such excesses. With an eye on his compatriots, who suddenly turned from worshippers of Germany into its ardent vilifiers, Duhem resolved, as was recalled by his daughter, "to say something good about the *Boches*" (the unflattering French nickname for Germans). In fact he urged his compatriots, too keen on the intuitive *esprit de finesse,* never to neglect developing the *esprit géométrique* or the laborious search for facts and their painstaking verification in which the Germans excelled. Duhem saw the basic fault of the German mind, eager to march forward systematically in every field, in the oversight of the crucial importance of the first step, or the very first link in the chain of any discourse. For unless that first link was reliable, the rest of the reasoning could not be safely anchored. The special function of the *esprit de finesse* was to ascertain intuitively the correctness of the first step or the soundness of basic primordial notions on which all reasoning depended.

The *esprit de finesse* and the *bon sens,* so often celebrated from Descartes on by French authors (above all by Pascal) were one and the same thing for Duhem. Yet even in that respect Duhem was not simply a Pascalian, that is, an intuitionist ready to be engulfed by the dictates of the "heart". He was a sober realist, never to be swayed, however slightly, by German idealism, much in vogue in the philosophical circles of the Third Republic. He was also the kind of realist, firmly committed to man's ability to seize immediately the reality external to him and the ontological order embodied in it, who needed no laborious demonstration to lay bare the dichotomic mind of the typical German professor. He was, in Duhem's succinct and graphic portrayal, lost in abstractions in the classroom and equally ready to be lost in the reality of "beer, pipe, and sauerkraut" at home.

Duhem could make short shrift not only of Kant, but also of Darwin. Not that Duhem was not an evolutionist. Long before it became a fashion, if not an obsession, of historians and philosophers of science to cast scientific history in the mould of the struggle of ideas for survival, Duhem did it, but he never reified mere concepts and pleasing similes, let alone mere words. As a realist he knew the limits of analogies and similes. Above all, he knew how to remain independent of prevailing fashions. His reference in the second lecture, dealing with the natural sciences, to Henri Fabre, the greatest entomologist of all time, is worth recalling for more than one reason. For Duhem the realist, facts were facts against which no theorizing could be offered as a refutation. Thus as long as Darwinism, or rather the mechanism of natural selection, was offered as an all-purpose explanation of the evolutionary process, even a few facts, to say nothing of the large number of facts marshalled by Fabre were a round refutation of Darwinism. Duhem indeed had such a high regard for Fabre as to convey to the Académie des Sciences his unwillingness to accept his impending election if thereby Fabre, 90 years old in 1913, was to be deprived of the honor. It is doubtful that in modern times there has been another academician with that measure of self-effacement.

Lecture II on the historical sciences throws light on the extent to which Duhem, the historian of science, was influenced by the painstaking reading, as urged by Fustel de Coulanges, of *all* documents available on one's subject. Actually Duhem's keen interest in history goes back to his years at the Collège Stanislas (1872–82). There the influence of Louis Cons, the foremost French author of history textbooks in the 1870s, almost made young Duhem choose history as his professional field. Curiously, in that lecture Duhem did not take up a favorite theme of his since 1906 when he first stumbled on the commentaries of Buridan and Oresme on Aristotle's writings. Those two teachers at the fourteenth-century Sorbonne were for him major glories of French learning, forerunners of Copernicus and Galileo in creating classical physics.

Such an interpretation of the origins of modern science was just as startling as was Duhem's attitude to what was *modern* science after 1900, namely, relativity theory and its chief support, non-Euclidean geometries. To a present-day reader Duhem's words on Riemann, Minkowski, and Einstein may seem outrageous if not simply ridiculous. After all, has not relativity theory become a most successful and indispensable tool all across the vast realm of physical science, ranging from atoms to galaxies? Duhem's disparaging words should, however, be seen in Duhem's own vision of what physical theory was ultimately about. As already noted, he did not see in the making of discoveries the chief aim of physics. The almost magically heuristic ability of mathematical physics to predict new facts had but a minor role in his *Théorie physique*. He certainly was not the kind of Frenchman who is upset by the rarity of French scientists among recipients of Nobel Prizes. For Duhem the chief glory of science was a secure systematization of all physical laws. By security he meant not only an inner con-sistency of the system, but also its reliable connection at every step with physical reality to which common sense or *bon sens* gave the indispensable access. Such is the basis of Duhem's diffidence about a relativity theory in which the impossibility of measuring the simultaneity of two events is turned into a denial of the possibility of ontological simultaneity. For Duhem clearly saw that almost from its inception relativity theory had taken on a pseudo-metaphysical if not anti-ontological message. In that development, which Duhem was one of the first to note, he rightly saw a deadly thrust at science itself.

Had he lived a decade or two longer, Duhem, the consummate logician, would not have spared that hapless development in which relativity theory first encouraged the relativization of the absolute and then the absolutization of the relative. Duhem's concern for the dictates of logic would have been no less justified by what happened to quantum theory, whose chief architects heedlessly rushed from the operational impossibility of measuring certain interactions exactly to the claim that therefore those interactions could not take place exactly in the non-operational or ontological sense. While the devotees of the Copenhagen interpretation of quantum mechanics have yet to wake up from their philosophical slumber, Einstein began to realize, only a decade or so after Duhem's death, that his relativity theory implied a realist ontology pointing to the absolute. Today, it no longer passes for a reckless attitude to point out what Eddington had noted already in 1920, that the expansion of the universe is an absolute reference system.

That Einstein himself, in the late 1940s, warned such a prominent logical positivist as Carnap that no physics would ever cope with man's sensing the *now*, may help one understand Duhem's real concern about relativity theory. That it was a concern of the *esprit de finesse* can best be seen from the elaborate disagreement that took place at the Sorbonne in April 1922 between Einstein and Bergson in the presence of a vast audience representing the elite of French intellect. No one could press Einstein more effectively on the question of time, which in relativity theory turned into a mere fourth dimension, than Bergson, whose philosophy centered on the still unexhausted richness of man's immediate experience of time, including his experience of the *now*. In the audience was Maritain, then 38 and a friend of Bergson, who in 1931 recalled that Einstein, in order to cope with Bergson's objections, invariably referred to his speaking of time only inasmuch as he was a physicist. But it was precisely that dichotomy between physics and ontology which was ultimately impossible according to Duhem. To be sure, he held that physics was independent of metaphysics in the sense that no specific physical theory could be constructed from basic metaphysical notions, the sole ties of physics with reality. Any deprecating of those ties was in Duhem's eyes the undermining of the very meaning of physical science as something relating to that reality which is physics.

Duhem's aim in these lectures was not to explain himself on every point, a task for which not even a dozen lectures would have been enough. Nor was his aim to be attentive to any and all exceptions to his broad generalizations. They were telling in spite of the fact that, for instance, such prominent German mathematicians as Hilbert and Klein stressed the importance of intuition. Duhem's aim was to strengthen French confidence in the French soul at a time when that soul seemed to falter. To his great credit Duhem went about his task with no touch of that chauvinism which, let us not forget, became a common word on account of the behavior of Nicolas Chauvin of Napoleonic times, who is described in *Larousse* as a "fanatical patriot". Duhem did not list a single Frenchman among those geniuses who like Newton, Gauss, and Helmholz were, according to him, able to rise above the limitations of national spirit.

Throughout those four lectures and the two articles added to them, Duhem is an example of moderation and humaneness which are carried to the reader by a style that made Duhem a master of French prose. Duhem is eminently readable even today and his message is no less relevant. The English-speaking world should be grateful to the translator who has once again succeeded in the delicate task of balancing two almost incompatible requirements: the exact rendering of the thought of a French author with the English reader's need for an idiomatic presentation of it.[2]

The reader may have two particular thoughts on reading this volume. One thought would be about still another book that would have been written by Duhem had his death not intervened. In the summer of 1916 he was proofreading the fifth volume of his immortal magnum opus, the *Système du monde*. As it turned out, the remaining five volumes were ready in manuscript, a heroic achievement of a mere nine years (1908–16). Indeed, by the summer of 1918 he might have been free to devote two months to writing a 300-page-long summary of that enormous work, bursting with information, a good part of it previously unpublished, on the history of cosmology from Plato to Copernicus and the various particular sciences pertaining to it. The marvellous readability of Duhem's *German Science* may suggest the masterpiece that might have come out of Duhem's pen in the form of a summary of the *Système du monde* aimed at the vast readership of non-specialists.

The other thought relates to the immediate reception of *German Science*. It was certainly received with the deepest appreciation by those for whom it was intended, namely, the members of the Student Association slated for active military service and for its members already on the battlefield. Duhem sent a copy to all of them. The notes of thanks he received he viewed as the highest recompense for his labors. He also sent copies, almost a hundred, to the large circle of those who were fully appreciative of his scholarly eminence. Among those who sent letters of heartfeld thanks were Léon Brunschvicg, Emile Boutroux, Tullio Levi-Cività, and Vito Volterra. Needless to say, letters of thanks poured in from French intellectuals who were also his personal friends, such as Bouasse, Chevrillon, Delbos, Fliche, Jordan, Jullian, Marchis, d'Ocagne, Violle, and others. A serious inquiry was made from Basel about translating the book into German. Houllevigue (professor of physics at the University of Montpellier and a former classmate of Duhem at the École Normale) brought the book, which sold out in two months, to the attention of many with a long review in the *Journal des Débats*.

All that amounted to not much in the way of response. But Duhem was never interested in immediate success. In practically everything he did, he took the long view. He trusted in the ultimate triumph of logic, the mainstay of truth in his eyes. A recently deceased author, who anticipated the prospect of having at the time of his death one reader, ten years later ten readers, and a century later one hundred readers, expressed Duhem's expectations well. These have been more than fulfilled: the past 40 years have not only witnessed the publication in the 1950s of the posthumous volumes of the *Système du monde*, but also the republication of its ten vast volumes. Several of the volumes have already seen a reprinting since. Duhem's three volumes of Leonardo studies, first published in 1906–13, were reprinted in 1955 and again in 1984. His *Théorie physique*, reprinted for the second time in 1981, found a world-wide readership through being translated into English in 1954 and made available in paperback in 1965. Two of Duhem's lesser-known books have been issued in English translation during the past dozen or so years[3] and more are possibly to come in addition to this translation. Plasma physicists are eagerly studying the reprinting in 1961 of his studies on hydrodynamics. Duhem's thought is very much at the center of interest in three different fields, a rare distinction for a scholar deceased three quarters of a century ago.

Not that Duhem was the kind of scholar whose chief interest lay in being personally remembered. His favorite simile was the woman whose image decorates French coins, generously throwing the seed to the four corners of the Earth. He saw the chief mission and glory of his beloved country in that selfless enrichment of mankind. Originality, profundity, and enduring vitality are the hallmarks of Duhem's contribution to that mission.

German Science

La science allemande (Paris: Librairie Scientifique A. Hermann et Fils, 1915).

These four lectures on *German Science* were
given at Bordeaux, under the auspices of
the Catholic Students' Association of the
University, on February 25, March 4, March 11
and March 18, 1915.

I dedicate these lectures to
THE CATHOLIC STUDENTS OF
THE UNIVERSITY OF BORDEAUX,
who requested and sponsored them.

With the help of God, may these humble
pages protect and promote in them
and in all their colleagues
the insightful genius of our France!

THE SCIENCES OF REASONING

LADIES AND GENTLEMEN:

If ever the word 'conspire' could be used in its fullest sense, it is assuredly used of the France that quickens under our eyes. Every breast breathes in unison, every heart throbs to the same feelings. One single soul animates this vast body of France. To save and redeem the soil of France, dear students, your elders and your fellow students are soaking it with blood beyond price. A short time ago I shook the hands of those of you who belong to the class of 1915. When I said to them, "Farewell! May God protect you!" I saw their eyes glisten with a flash of joy. A young Frenchman is only completely happy in fulfilling his duty when it is very dangerous. And I sometimes see you, the younger fellows of those students, clench your fists, for you dream of vengeance and believe that you already possess avenging arms. All about you mothers, wives, sisters, daughters of soldiers strive to outdo each other in alleviating the afflictions of the combatants or the sufferings of the wounded. And if it be that some brows wear the veil of sorrow, their radiance appears to us to transfigure their mourning, for the acceptance of sacrifice casts its halo there.

In the midst of this 'conspiracy', he who is about to speak
to you felt profound anguish. Except through prayer he was
incapable of taking part in the great common task. M. l'Abbé
Bergereau had compassion on this sorrow, brought about by a
sense of uselessness. He said to me: It isn't the soil of France
alone that has been invaded. Foreign thought has taken French
thought captive. Go sound the charge to deliver the soul of
the Fatherland!

Assigned my battle station, I come running. The post is
without danger, and will therefore be without glory. I shall
have no occasion to pour out my blood there, but I shall pour
out all the devotion that my heart contains.

I come before you to take my humble part in the national
defense.

Everyone has studied, to some extent, the principles of
arithmetic or geometry. It is therefore common knowledge that
the propositions of which these sciences are composed are
divided into two categories: A few axioms comprise one category;
innumerable theorems make up the other. Of the sciences of
reasoning, arithmetic and geometry are the most simple and,
consequently, the most completely perfected. In each of the latter
sciences we ought likewise to distinguish theorems from axioms.

Axioms are the sources, the principles, of theorems. Deductive reasoning proceeds by rules which, for the human mind, are
like spontaneous effects of natural instinct but analyzed and formulated by logic, obliging whoever admits the truth of the axioms
to accept equally the theorems which are their consequences.

What is the source of the axioms? We ordinarily say that they
are drawn from common knowledge, that is to say, that every
sane man holds their truth to be certain before studying the
science whose foundations they will become. Let a man, for
example, or a child in possession of his reason but still ignorant
of arithmetic and geometry, hear these propositions formulated:

The sum of two numbers is invariable when one inverts
the order in which they were added to each other.

A whole is greater than each of its parts.

Through two points, a straight line can always be drawn, and no more than one can be drawn.

As soon as this man or this child had directed his attention to the proposition which he has just heard, has fixed it with the mind's eye (*intuere*), he will hold it as true. Thus we say he has an *intuitive* certitude of it.

It is not the same with a theorem. Someone who has not studied arithmetic will not know whether he is confronted with an error or a truth when he hears this proposition formulated: The least common multiple of two numbers is the quotient of their product divided by their greatest common factor. His uncertainty will be the same if he does not know geometry and we say to him that the measure of the volume of a sphere is equal to the product of its surface multiplied by one-third of the radius. In order for him to come to regard these propositions as quite certain truths, he must patiently pass through (*discurrere*) a long sequence of reasoning which will show him one step at a time how the certainty possessed by the axioms is transmitted to the theorems. This is why we have only a *discursive* knowledge of the truth of theorems.

To indicate the immediately obvious character of axiomatic evidence, we readily compare its obviousness to perception: we say that *we see* that such a proposition is true. Its certitude is *palpable*. The faculty by which we know the axioms is given the name of 'sense': it is *common sense*, or *good sense*.

Often, also, to better distinguish the immediacy of the intellectual operation which recognizes the truth of a principle from the deliberativeness of the discursive reasoning suited to the demonstration of theorems, we give to the former operation the name of 'feeling'. It is the *feeling for the truth*. We feel the truth immediately when our attention falls upon a principle, as the sight of a masterpiece of art makes us immediately experience the feeling of beauty, or the account of a heroic act makes us immediately experience the feeling of goodness.

"We know the truth not only by reason, but also by the heart," said Pascal, "and it is through the latter sort of knowledge that we know first principles." "And it is upon such knowledge of the heart and of instinct that reason must rest, and base all its discourse. The heart feels that there are three dimensions in space, and that numbers are infinite. Reason then demonstrates that there are no two squared numbers one of which is twice the other. Principles are intuited, propositions are inferred [*Les principes se sentent, les propositions se concluent*]; all lead to certitude, though by different routes."[1]

Good sense, which Pascal calls the "heart" here, for the intuitive perception of the obviousness of the axioms, and the deductive method to arrive by the rigorous but slow progress *of discourse* at the demonstration of the theorems: there we have the two means that human intelligence uses when it wishes to construct a science of reasoning.

But all minds are not equally adapted to the use of each of these two means.

It is not necessary to have progressed far in one's study of arithmetic or geometry in order to discover how arduous it is to handle the rigorous reasoning by means of which these two sciences proceed. A number of people, and they are not stupid, cannot conform their intelligence to this minutely prudent and severely disciplined approach. Beginners, minds in rebellion against mathematics, are not alone, moreover, in coming across redoubtable difficulties in the use of the deductive method. The most skillful algebraists and the most illustrious geometers have been seen to stumble over the same difficulties. The great men who, from the seventeenth century through the middle of the nineteenth century, created algebra, integral calculus, and celestial mechanics, often justified their most important discoveries by means of defective processes of reasoning, or even by flagrant paralogisms. One of the essential tasks which mathematicians of the nineteenth century accomplished under the inspired prompting of Augustin Cauchy was to take up again the entire work of their predecessors in order to complete and set right their processes of reasoning, and show them "how they ought to have discovered what they had so felicitously invented".

What is the most frequent and most dangerous cause of these errors, to which deductions of the most skillful mathematicians are so often and so easily exposed? Jumping to conclusions.

Have you ever had the experience of coming down from a mountain on a steep and slippery footpath behind a mule? Have you observed the precautions with which the mule puts one foot forward only after having carefully secured the other three? Have you noticed the caution with which it tests with the tip of its shoe the solidity of the rock where its fourth foot will be placed? Irritated by this prudent but annoying slowness, haven't you taken advantage of the first widening of the footpath to overtake the tiresome animal? Deductive reasoning proceeds in this mulish fashion. It advances no proposition until it has rigorously demonstrated all the preceding ones; and the new proposition will not be established with any less care.

The discoverer [*inventeur*] who has intuited [*deviné*] some truth impatiently chafes at the tedium and the detailed precautions it is necessary to take in order to give complete certitude to his discovery. From time to time he happens to bypass some intermediate step which he judges to be of little importance and easy to supply. A dangerous haste! It is almost always a leap of this sort which causes him to slip and fall into error.

When Laplace, at the end of a process of reasoning, came to a conclusion which, in another connection, he knew was wrong, and wished to discover at what point his deduction was faulty, he went back up his chain of reasoning to the spot at which he read such words as these: 'We readily see that . . .'. Every time, the unintentionally fallacious reasoning lay in the intermediate steps which the great astronomer had believed he could skip.

This slow and prudent procedure of the deductive method, which only advances one step at a time, each of whose forward movements must obey the rigorous discipline imposed on it by the rules of logic, is above all else the style which suits the German intellect. The German is patient. He knows nothing of feverish precipitation. Consequently, there certainly are more intellects in Germany than elsewhere capable of forging a long chain of reasoning each of whose links has been minutely tested.

Mathematicians, we have pointed out, accomplished during the second half of the nineteenth century the arduous task of proving anew through impeccably rigorous processes of reasoning many a theory which their predecessors had too hastily formulated. As we have also pointed out, it was a French mathematician, Augustin Cauchy, who first recognized the necessity of such a task and showed, by his example, how it ought to be accomplished. That task was carried on in many different lands. To the Norwegian Henrik Niels Abel it owes one of its essential tools, the notion of the uniform convergence of a series. But if there is one school which has, so to speak, made this work its specialty, it is assuredly the school directed by the algebraist Weierstrass in Berlin. Weierstrass was a past master in the art of discovering faults of reasoning where his predecessors believed they had produced a deduction of irreproachable rigor. With consummate skill he replaced the defective parts of the chain of reasoning with new links which no longer risked the least discontinuity. The disciples of Weierstrass inherit the logical strictness of their master. One of them, Professor Hermann Armandus Schwartz, likes to say: "I am the only mathematician who has never made a mistake." Schwartz, it is true, purchases this impeccable security at the price of extreme minuteness. In the course of his deductions, he never gives the reader the trouble of supplying the smallest intermediate step. One of my friends, today among our great geometers, who once took Schwartz's course of lectures at Göttingen, told me to how hard a trial the deliberateness of the German geometer subjected his French nerves.

This great aptitude for deduction with impeccable rigor is, we believe, the mark of the German intellect. This is what will impress on German science its characteristic qualities. This is what will distinguish German science from theories [*doctrines*] developed in France, Italy, and England. It will explain both the good qualities and the shortcomings of the methods in favor beyond the Rhine.

There are, among men, elite intellects in which each faculty is very fully developed and yet maintained in the most harmonious accord and perfect equilibrium with respect to every other faculty. But minds thus happily constructed are quite rare. An organ of the body can scarcely undergo exceptional development without weakening and diminishing the vitality of neighboring organs. The same thing holds for the mind. The extreme vigor of one faculty is often paid for by the enfeeblement of another. Those whose lively good sense allows them to seize upon the truth through an intuition as quick as it is accurate are sometimes also those who have the hardest time submitting themselves to the prudent discipline and rigorous deliberateness of the deductive method. On the other hand, those who follow most minutely the rules of the deductive method frequently fail through lack of common sense.

Beneath the guise of philosophical gravity, Descartes frequently hides the caustic wit of a man of pitiless irony. It is assuredly this wit which caused him to write at the beginning of the *Discourse on Method:*

> Good sense is of all things in the world the most equally distributed, for everybody thinks himself so abundantly provided with it, that even those most difficult to please in all other matters do not commonly desire more of it than they already possess. It is unlikely that this is an error on their part; it seems rather to be evidence in support of the view that the power of forming a good judgment and of distinguishing the true from the false, which is what is called Good Sense or Reason, is naturally equal in all men.[2]

No, it is not true that the ability to discern intuitively the true from the false, that is, good sense, is equally developed in all men. Do we not constantly say that a particular man has good sense, while another is bereft of it? And do we not quite often find this absence of common sense, of good sense, among those people who have great facility in setting forth a long series of deductions? Their mind is that of Chrysalis: "reasoning banishes reason".

If great ability to follow the deductive method frequently has as its counterpart mediocrity of intuition, it will appear natural to us that the Germans, often so skillful in linking together long and rigorous chains of reasoning, might also often be ill-furnished with good sense, and that in numerous cases the latter fault, like the former skill, will characterize the products of their intellect.

When a man is strongly endowed with a physical or intellectual faculty, such a person experiences a lively enjoyment in using it. On the other hand, it is tiresome for him to bring into play an underdeveloped organ or a mediocre aptitude. The German, then, quite skillful in the use of the deductive method, but weakly endowed with intuitive knowledge, will multiply the occasions on which the former may be applied, and restrict as much as possible the circumstances which call for the latter.

A very common fault among geometers has always been that of seeking out occasions for exercising their aptitude for reasoning and showing their skill in the art of linking together syllogisms. The *Logic of Port Royal*, inspired by Descartes and Pascal, reproached geometers with the failing of trying "to prove things which have no need of proof".

> Geometers admit that it is not necessary to attempt to prove that which is self-evident. They often do so nevertheless, because, being more given to convincing the mind than to enlightening it, as we were saying, they believe that they will convince it better by finding some proof of

even the most obvious things than by simply proposing them, and letting the mind recognize their obviousness.

This is what brought Euclid to prove that the two sides of a triangle taken together are greater than one alone, although that is quite obvious from the notion of the straight line alone, which is the shortest line which can be drawn between two points.[3]

What might Descartes, Pascal, and the authors of the Port-Royal *Logic* have said if they had come to know the unbelievable excess to which the desire to prove everything has led certain mathematicians of the contemporary German school? Weierstrass and his better disciples applied themselves to taking truly useful precautions against algebraic demonstrations which were insufficient or inexact. The work which they did was indispensable, unless mathematics were to become the mistress of error. Those who have come after them, finding that the genuine shortcomings of former processes of reasoning had been remedied, worked incessantly at correcting imaginary faults, at filling in lacunae that every properly constituted mind recognized as inoffensive abbreviations, in detailing trifles—in a word, in splitting hairs. With them, the discourse of mathematics has become so complicated, so cramped with captious rules, that one no longer dares speak it for fear of lacking rigor. By incessantly refining the recognized processes of reasoning, these logicians destroy the desire for discovery and the ability to make it in themselves and in their disciples. Yet one also finds, even in Germany, geometricians such as Felix Klein, who assert the rights of the spirit of invention against this inordinate critique.

At the same time as he looks for occasions to exercise his argumentative faculty, the German mathematician avoids, as much as he can, circumstances which call for intuition. To that end, he restricts as far as possible the number of axioms he invokes.

Of all the sciences, that which best satisfies this tendency is algebra. All of algebra, in fact, is no more than an immense prolongation of arithmetic, and the deductive method is all that is necessary to produce this prodigious development. The only axioms which algebra needs are thus the axioms on which arithmetic is based, that is to say, a quite small number of extremely simple and glaringly obvious propositions concerning the addition of whole numbers. We should not be surprised that the German mind has passionately and successfully given itself over to algebra.

But the German mathematicians are not content with being skillful algebraists. That science, wherein the role of axioms is so reduced, wherein the deductive method suffices for all development, is so well adapted to the form of their intelligence that they do all they can to dissolve all other mathematical sciences into it. They would like geometry, mechanics, and physics to be no more than chapters in algebra. How they have gone about transforming this wish into reality bit by bit, and the disadvantages resulting therefrom for the sciences thus reduced to algebra, we have set forth elsewhere.[4] To avoid an over-technical presentation, we shall not repeat it here.

Not only does a man take pleasure in the exercise of a healthy and vital organ and find the exercise of an infirm one painful; he knows, furthermore, that he can count on the healthy one, while he distrusts the other.

The mind quite well-adapted to deduction, but poorly endowed with good sense, will give full faith to propositions demonstrated by the discursive method. It will readily doubt the propositions which intuition reveals to it. What are the consequences of these two tendencies? Are they not to be recognized with a particular clarity in the principal systems of German philosophy? We shall examine this matter briefly.

There are two sources of certitude: propositions receive their certitude from demonstration, and principles take theirs from common knowledge. The latter is not of a different value or kind from the former. Both are equally certain. To speak more precisely, we ought to say that there is a single source from which all certitude flows, the source that provides certitude to principles. For deduction creates no new certitude. All that it can do, when followed impeccably, is to carry over to the consequences the certitude which the premisses already possessed, without losing any of its force on the way. As Pascal said:

> Knowledge of first principles, such as *space, time, movement, numbers*, is as sure as any knowledge which reasoning gives us. It is upon these intuitive cognitions of the heart and instinct that reason must rest, and upon them it must base all of its discourse. . . . And it is as ridiculous for reason to demand proofs of its first principles from the heart before consenting to them, as it would be ridiculous for the heart to demand of reason that it provide emotional conviction for all the propositions which reason proposes before it was willing to admit them.[5]

It is precisely the first of these two absurdities that will characterize those whose aptitude for deduction has taken an exaggerated development at the expense of common sense. Too confident in the discursive method, too distrustful of intuition, they end by imagining that the discursive method alone gives to its conclusions a certitude of which intuition is incapable. As if the house could be any more solid than the foundations upon which it rests! Consequently, they hypnotize themselves with the illusion of a dream for which Pascal himself perhaps showed too much sympathy: they follow the chimera of an exclusively deductive method.

> This true method, which would fashion demonstrations of the highest excellence, if it were possible to arrive at them, would consist in two principal procedures: first, never to employ a word whose meaning one had not clearly explained previously; second, to advance no proposition which one could not demonstrate by means of previously known truths. That is to say, in a word, one would proceed by defining all terms and proving all propositions.[6]

Without doubt no one will proclaim or even admit to himself that he is attempting to develop such a method. It is too clear that such a development is senseless, that the first definitions will always be composed of undefined terms and the initial demonstrations constructed with the aid of undemonstrated propositions. But one will act, at least, as if this method were an ideal which it were desirable constantly to approximate, though certain that it can never be attained. In that case, one would push on continually, constantly dig deeper, in order to define the terms used up to then without definition, and to demonstrate the propositions previously accepted as principles.

Such disappointing research, since it could never satisfy our desire to know, could never come to rest.

> Whatever end to which we think to attach ourselves and by which we might bolster ourselves, it comes loose and departs from us. And if we follow it, it escapes from our grasp, eludes us, and leads us on an endless chase. Nothing holds still for us . . . . we burn with the desire of finding a stable position and an ultimately unchanging foundation, in order to erect thereon a tower rising into the infinite. But all our foundations crack, and chasms open up to the very depths of the earth.[7]

He, then, who locates the principle of certitude in discursive reasoning instead of in the intuitive knowledge which derives from common sense, cannot but land in the absolute skepticism that calls all propositions into question.

There is only one way of avoiding this despair of the intellect, and that is to hold firmly that demonstration never creates certitude; that, whether mediately or immediately, all assurance of the truth comes to us through good sense. Thus Pascal, who constantly meditated on the issue, said: "We have an impotence to prove which cannot be overcome by any dogmatism; we have an idea of the truth which cannot be overcome by any pyrrhonism."[8]

Highly skilled at deduction, the German mind is poorly endowed with common sense. It has a limitless confidence in the discursive method, whereas its confused intuition gives it only a weak assurance of the truth. It is consequently peculiarly vulnerable to slipping into skepticism. It has frequently and ponderously fallen into it; Kant vigorously pushed it there.

What is the *Critique of Pure Reason*? The longest, the most obscure, the most confused, the most pedantic commentary on these words of Pascal: "We have an impotence to prove which cannot be overcome by any dogmatism." Minds too exclusively deductive, such as Descartes's or Spinoza's, believed that syllogisms would allow them to establish beyond a doubt the first principles of metaphysics and morals. More narrowly constrained by the rigorous discipline of the discursive method, Kant applied himself to showing that they sinned against the rules of logic, that their syllogisms were not conclusive, and therefore that doubt is the only legitimate conclusion of "pure reason".

Assuredly, absolute skepticism is not the last word of the philosopher of Königsberg. Kant desires equally to give proof of the second part of Pascal's formulation, to show that "we have an idea of the truth which cannot be overcome by any pyrrhonism." This is the object of *The Critique of Practical Reason*. But does this idea of truth, against which the assaults of skepticism ought to break, possess sufficient breadth and depth to support the entirety of our knowledge? Not at all. It restricts the scope of knowledge. It makes knowledge just broad enough to serve as the basis of morals. It reduces knowledge to affirming for us the imperative character of obligation. Even within these narrow limits it does not allow for the possession of that perfect certitude vainly demanded of pure reason. The certitude that it enjoys is of another order, and, so to speak, of an inferior quality. It is capable of directing our acts, but not of satisfying our reason. It is but a practical certitude. Let us listen, for example, to the terms in which Kant speaks of the existence of God:

> Since there are practical laws (moral laws) which are
> absolutely necessary, if these laws necessarily presuppose
> some existence as the condition of the possibility of their
> *obligatory* force, it must be that that existence is a *postulate*,
> because in effect the conditioned [*conditionné*] from which

the reasoning proceeds in order to arrive at this deter-
minate condition is itself known *a priori* as absolutely
necessary. We shall show later on, on the the subject of
moral laws, that they not only presuppose the existence of
a Supreme Being, but further, since they are absolutely
necessary from another point of view, that they postulate
it rightly; a postulate, in truth, only practical.[9]

This certitude of restricted and inferior quality, this purely
practical certitude which preserves only the essential principles
of morality—does Kant at least attribute it directly to the intuitions of common sense? He is too much enamored of the deductive method to proceed in this fashion, too deeply imbued with
the methods of reasoning customary with geometricians. This
possibility of principles of practice *a priori* he only dares to propose, at the beginning of his *Critique of Pure Reason,* as the
sequel to a long demonstration, wherein definitions, theorems,
corollaries, scholia, and problems are brought together as in a
treatise on algebra.

The disciples of Kant have gone further than their master.
They have relinquished practical certitude, the last remembrance of the evidence which common sense confers. They
have admitted no assurance of truth to be allowable, moreover,
other than that of pure reason, that is to say, of discursive
knowledge. And since that, by itself, is incapable of giving
what one expects, they have ended with complete idealism,
absolute skepticism.

Excessive confidence in deductive reasoning, distrust and
disdain for the intuitions which common sense supplies—these
then are the causes which necessarily produce idealism and
skepticism. They also engender sophistry.

If common sense is not the certain source of truth, then
what good is it to take our axioms from it? Aren't we able just
as well to invent them at our good pleasure? Provided we set
out a long chain of conclusive syllogisms from postulates
freely posed, our reason, which has a penchant for deduction

alone, should find itself fully satisfied. Its very contentment will be in proportion to its ability to deduce rigorously without any recourse to good sense.

Who of us has not known men experienced in all the intricacies of reasoning, mathematicians for example, for whom it was lively enjoyment to follow quite logically and quite extensively the consequences of a highly disconcerting paradox?

At an early period, German thought gave evidence of this strange and dangerous fault, which is sophistry. It then found enjoyment in constructing vast systems of thought upon postulates which had no common sense in them.

Nicolas Crypfs was born in 1401 at Cues, on the Moselle, the son of a simple fisherman. After having studied at Heidelberg, and at Padua, where he took a doctorate in law in 1424, he returned to Germany. In 1431, as archdeacon of Liège, he attended the Council of Basle. He was among the members of the Council who remained faithful to the Pope. Eugenius IV, Nicolas V, and Pius II entrusted him with important missions. In 1448 Nicolas V named him cardinal-priest with St. Peter-in-Chains as his titular church. A German cardinal was at that time, as one historian says, as rare as a white crow. Consequently, Nicolas of Cusa was often called *Cardinalis teutonicus.* In 1450 he was promoted by Nicholas V to the bishopric of Brixen in the Tyrol. He died at Todi, in Umbria, on August 11, 1464.

A person of considerable importance in the Church, Nicolas of Cusa was a man of science. He presented to the Council of Basle a project for reforming the calendar. He was a geometer. We also owe to him an attempt at squaring the circle which is not without ingenuity.

Germans readily and rightly hail Nicolas of Cusa as the first truly original thinker to be produced by Germany. The German cardinal, indeed, in his basic work, *On Learned Ignorance* (*De docta ignorantia*) constructed an entire metaphysics, quite thoroughly impregnated by Neo-Platonism, to which his later writings provided numerous additions. Now, that metaphysics, developed with the true virtuosity of a dialectician, rests entirely upon the following axiom, which common sense declares to be a formal contradiction: In every order of things, the maximum is identical with the minimum.

Before the middle of the fifteenth century, then, the tendency to sophistry concealed in the German mind had shown what it was capable of. Many times since then this tendency has directed the construction of the metaphysical systems which have been so characteristic of German philosophy. Of these disconcerting endeavors, we content ourselves with mentioning the strangest, perhaps, and at the same time the most celebrated. Up till our own time, in thought from beyond the Rhine, it has exercised as much if not more influence than Kant's critique. I refer to the doctrine of Hegel. Could one more harshly and more insolently trample underfoot the first principles of common sense than the Hegelian metaphysics does? In that metaphysics, as a matter of fact, the essential axiom, quite similar to that of Nicolas of Cusa, is the following:

In every order of things, contradictories are identical, for the *thesis* and the *antithesis* together make but one entity in the *synthesis*, which is the truth.

What ought to be noted here is not that a Hegel should be found among the Germans. Among all peoples and at all times one comes across wretched maniacs who reason from absurd principles to their ultimate conclusions. What is serious in the present situation is that the German universities, rather than holding Hegelianism to be the delirium of a mad dreamer, hailed it with enthusiasm as a doctrine whose splendor eclipsed all the philosophies of Plato or Aristotle, Descartes or Leibniz.

The excessive taste for the deductive method, the disdain for common sense, have truly made German thought just like the house of Chrysalis: reasoning, therein, has banished reason.[10]

LECTURE II:
THE EXPERIMENTAL SCIENCES

LADIES AND GENTLEMEN:

In algebra and geometry, and also in metaphysics when it is soundly constructed, the axioms are utterly simple. Once our attention is fixed upon any one of them, immediately the sense of it becomes perfectly evident to us, and its certitude fully warranted. Consequently, in these sciences, "the principles are palpable, but removed from ordinary use; so that for want of habit, it is difficult to turn one's mind in that direction: but if one turns it thither ever so little, one sees the principles fully, and one must have a quite inaccurate mind who reasons wrongly from principles so plain that it is almost impossible they should escape notice."[1]

Things are quite different with the experimental sciences.

In these sciences, the principles are not called axioms any more but rather hypotheses or suppositions, words which must be understood in their etymological sense, namely, foundations. They are also called experimental laws, or truths of observation. Simply focussing attention upon the statement of a hypothesis or a law does not in any way allow us to hold it to be true. It would be licit to assent to it only after the complex and prolonged labor of experiment which tests it.

How could one educe from experiment a hypothesis appropriate to play the role of principle in an observational science?

At the time when I was an assistant in a physics laboratory, I chatted daily with several friends who were assistants in the neighboring laboratory. This was the laboratory of Louis Pasteur, and at that time he was experimenting there with the first rabies vaccine. My friends told me of the manner in which "the boss" worked, and I have the most vivid remembrance of their accounts.

Pasteur arrived at the laboratory having in his head what Claude Bernard called a preconceived idea, that is, a proposition which he wished to submit to experimental verification. His assistants, under his direction, prepared experiments which, according to that preconceived idea, ought to produce certain results. Most of the time the expected results were not those which the experiment in fact produced. The experiments were then repeated from the beginning, and with greater care. Another failure. A third attempt was then launched, only to result in a further failure. My friends the laboratory assistants were often astonished at the obstinacy of "the boss", intoxicated by the pursuit of the consequences of what was obviously an erroneous preconception. Finally a day came when Pasteur announced an idea different from the one which experiment condemned. One then realized with admiration that none of the contradictions to which the latter had led had been in vain; for each of them had been taken into account in the formation of the new hypothesis. It, in its turn, had its corollaries put to the test of facts, and, quite often, the process resulted in new failures. However, in the process of destroying this hypothesis, these failures prepared the conception of a new idea. So, little by little, through this sort of struggle between preconceived ideas that suggested experiments, and the experiments that constrained the preconceived ideas to be transformed, a hypothesis was formed which conformed perfectly with the facts and was capable of being received as a new law of physiology. In this work of successive improvements upon an initial necessarily rash and often false idea which finally leads to a fruitful hypothesis, the deductive method and intuition each play their role. But how much more complex and difficult it is to define their roles than it is in a science of reasoning!

In order to draw out of the preconceived idea consequences that can be compared with facts, which experimental proof will either confirm or condemn, one must deduce. Such deduction is often a quite long and delicate process. It is essential that it be a rigorous process, under pain of making the observational testing depend upon propositions which could not be derived from the hypothesis, and thus of rendering this testing illusory. This reasoning, however, cannot in general be conducted *more geometrico,* under the form of a series of theorems. The proposition whose consequences one wishes to deduce would not lend itself to this process. The ideas on which it depends are no longer highly abstract but quite simple concepts, like the first objects of the mathematical sciences, or ideas made in a well-known fashion by definition using those concepts. These are ideas richer in content but less precise, less analyzed; they issue more immediately from observations. To reason exactly with such ideas, the rules of syllogistic logic are not adequate. They must be assisted by a certain sense of soundness that is one of the forms of good sense.

In another way, again, good sense will intervene at the moment at which one realizes that the consequences of a preconceived idea are either contradicted or confirmed by the experiment. This realization is in fact far from being entirely simple; the confirmation or contradiction is not always explicit and straightforward, like a simple 'yes' or 'no'. We stress this point a bit, for it is important.

From his preconceived idea an experimenter such as Louis Pasteur infers this consequence: if one injects one particular substance into rabbits, they will die. The control animals which have not had the injection will remain in good health. But this observer is aware that a rabbit might die, at times, from other causes than the injection whose effects are being studied. He knows equally that certain particularly resistant animals can sustain doses of an injection which would kill most of their fellows, or, further, that an inept administration can impair the injection and render it inoffensive. If, then, he sees one inoculated rabbit live or one control animal die, he need not conclude directly and overtly to the falsity of his preconceived idea. He could be faced with some accident of the experiment which need not require the abandonment of his idea. What will determine whether these failures are or are not of such a nature that the supposition in question must be renounced? Good sense. But this determination will be of just the same type as a judgment in a legal proceeding wherein each of the two parties is faced with evidence some of which tends toward conviction and some toward exculpation. Good sense will not return its verdict until after having weighed the pros and cons with mature consideration.

Guaranteeing the soundness of still-imprecise processes of reasoning by means of which consequences capable of being proved by experiment are drawn from a preconceived idea, and estimating whether such proof ought to be taken as indicative or not, do not exhaust the task which devolves upon good sense.

When the factual proof has turned against the preconceived idea, it is not enough simply to reject it. One must substitute for it a new supposition which has the possibility of standing up better to experimental testing. Here it is necessary (Pasteur excelled in doing this) to pay attention to what each of the observations which have condemned the initial idea suggests, to interpret each of the failures which destroyed that idea, and to synthesize all these lessons for the purpose of fabricating a new thought which will pass once again under the scrutiny of the actual results. What a delicate task, concerning which no precise rule can guide the mind! It is essentially a matter of insight and ingenuity! Truly, in order to perform this well, it is necessary that good sense should transcend itself, that is, push its strength and its suppleness to their very limits, that it become what Pascal called the intuitive mind [*esprit de finesse*].

Who can forget that admirable page of Pascal's in which he contrasted this intuitive mind to the mathematical mind [*l'esprit de géométrie*], skill at arrangement with the rigor of the deductive method?

In the mathematical mind, he said "one sees the principles fully, and one must have a quite inaccurate mind who reasons wrongly from principles so plain that it is almost impossible that they should escape notice."

> But in the intuitive mind the principles are found in common use, and are before the eyes of everybody. One has only to look, and no effort is necessary; it is only a question of good eyesight, but it must be good, for the principles are so subtle and so numerous, that it is almost impossible but that some escape notice. Now the omission of one principle leads to error; thus one must have very clear sight to see all the principles, and in the next place an accurate mind not to draw false deductions from known principles.
>
> All mathematicians would then be intuitive if they had clear sight, for they do not reason incorrectly from principles known to them. . . . But the reason that mathematicians are not intuitive is that they do not see what is before them, and that, accustomed to the exact and plain principles of mathematics, and not reasoning till they have well inspected and arranged their principles, they are lost in matters of intuition where the principles do not allow of such arrangement. They are scarcely seen; they are felt rather than seen; there is the greatest difficulty in making them felt by those who do not of themselves perceive them. These principles are so fine and so numerous that a very delicate and very clear sense is needed to perceive them, and to judge rightly and justly when they are perceived, without for the most part being able to demonstrate them in order as in mathematics; because the

principles are not known to us in the same way and because
it would be an endless matter to undertake it. We must
see the matter at once, at a glance, and not by a process
of reasoning, at least to a certain degree. And thus it is
rare that mathematicians are intuitive, and that men of
intuition are mathematicians, because mathematicians wish
to treat matters of intuition mathematically, and make
themselves ridiculous, wishing to begin with definitions
and then with axioms, which is not the way to proceed in
this kind of reasoning. Not that the mind does not do so,
but it does it tacitly, naturally and without technical rules;
for the expression of it is beyond all men, and only a few
can feel it.[2]

To set out from principles which are quite clear, the analysis
of which has been pushed to extremes, the content of which
has been described in its least detail; then to progress a step at
a time, patiently, minutely, in a manner disciplined with an
extreme severity by the rules of deductive logic—it is in this
that the German genius excels. The German mind is essentially
the mathematical mind. On the other hand, as we have seen, it
is wanting in good sense. How then could it possess that perfection of good sense which is the intuitive mind? No, do not
require of it that suppleness, acuity, delicacy which penetrate
and insinuate themselves into the obscure and complex recesses
of reality. Do not expect it to reason precisely and extensively
in the absence of definitions, because it is not really possible to
define the ideas which the perception of reality furnishes
immediately; or without syllogisms, because one cannot proceed from principles already formulated at the precise moment
that one wants to discover new principles; or without any other
guide or guarantee than a natural awareness of the true. The
German is a mathematician. He is not intuitive.

The German is not intuitive. Do you doubt it? Then, listen
to a confession of the fact. He who formulates it is, among
Germans, justly reputed for his exceptional intuitiveness. He is

in turn no less apt in the recognition of what is lacking in his fellow countrymen. "The great art of moving directly from comprehension to application," writes Prince von Bülow, "or even the greater talent of doing what is necessary in obedience to a sure, creative instinct without reflecting on it for a long time or racking one's brain—there is what we have lacked and what we shall lack on many occasions."[3]

The German is bereft of the intuitive mind.

Also, among all the great men who have set the foundations of experimental science from the seventeenth century right down to our own times, among the creators of physics, chemistry, biology, one comes across very few Germans.

However, since the middle of the nineteenth century, physics, chemistry, and biology have attracted the attention of a multitude of Germans who have brought about in each of these sciences very great progress. How can we reconcile this development of German experimental science, the scope and power of which are incontestable, with what we have just said? Let us try to show this.

In proportion as an experimental science is perfected, the hypotheses upon which it rests, at first hesitantly and confusedly, become more precise and firm. The numerous and varied tests to which the hypotheses have been put record the circumstances under which the facts have a high probability of corresponding to the consequences drawn from such hypotheses. From this moment on, these suppositions serve as principles in processes of reasoning at the conclusion of which is found the clarification of certain observations or the prediction of certain events. Science remains experimental in its essence. It always moves from propositions which experiment has suggested in order to conclude with propositions of which experiment alone can assure the truth. But science utilizes the deductive method more and more.

It even happens that further progress comes about this way. The initial hypotheses no longer rest exclusively upon ideas clearly conceived, but upon ideas which may be counted and measured, that is, upon quantities. Such suppositions then take the form of algebraic or geometrical propositions. It is no longer simply a question of deduction considered generally. It is to mathematical reasoning that experimental science has recourse in order to draw out from principles furnished by observation the consequences that observation will have to verify.

Successively, through the ages, we have seen various chapters of physical science taking this mathematical form. As early as the time of Plato and Aristotle, Eudoxus and Calippus strove to construct a mathematical theory by which it might be possible to save all that the senses establish concerning the motion of the heavens. Euclid already set forth in mathematical form the laws of the rectilinear propagation and reflection of light. With Archimedes, the statics of solid bodies and hydrostatics, in their turn, took on mathematical form. The Middle Ages were too little inclined to mathematics to do anything new in this way. They limited themselves to a qualitative analysis of the motion of projectiles and the fall of bodies, working out with great common sense the mathematical dynamics which was to usher in the work of Galileo, Descartes, Pierre Gassendi, and Torricelli. Today there is no longer any field of physical science in which one can reason without the constant assistance of algebra and geometry.

Of course, the continuous use of mathematical reasoning has not changed the experimental character of the sciences. Their hypotheses are not principles of which simple common sense makes us fully certain. The unique object of these suppositions is always to produce consequences which conform to reality. And when some disaccord erupts between the corollaries drawn from the suppositions and the results of observation; when in the judgment of good sense this disaccord is intolerable, then the hypotheses in question ought to be abandoned, to make way for new foundations. Just taking on a mathematical vesture does not give to a physical science its definitive investiture.

Has not the first such science to develop according to the laws of mathematics been thrown into utter confusion many times? Did not the astronomy of Eudoxus, which was devised using only rotations concentric to the Earth, capitulate before the astronomy of Ptolemy, which took into consideration eccentric and epicyclic rotations? In the opinion of the astronomer of Peluse [Ptolemy], all the movements of the heavens were centered about an immobile Earth. Did not Copernicus then substitute the fixity of the sun for the fixity of the Earth? Did not the astronomy of Kepler substitute for these combinations of circular movements a single elliptical movement? Finally, did not the astronomy of Newton renounce all these kinematic hypotheses for the sake of the dynamic hypothesis of universal gravity?

An experimental science does not acquire the unchangeable certitude of algebra or geometry, whose axioms are declared by common sense to be absolutely true, as a consequence of its general use of the deductive method or even of mathematical reasoning. At best, hypotheses which support it are, for a time, regarded as beyond contestation, and one expects with extreme probability that the corollaries that can be deduced from such hypotheses will agree with the facts which will be observed.

The intuitive mind, then, so necessary to the origins of science, takes but a small part in its development. It is necessary to have recourse to the mathematical mind in order to draw out all the consequences, even the most remote ones, from hypotheses which have been made explicit and consolidated by the intuitive mind. At that point, therefore, the German genius, which had been ill-equipped for the discovery of the principles of the science, finds itself marvellously apt for making them produce all the corollaries which they contain.

For the Germans, then, an experimental science is born the day that it takes deductive form, or better still, the day it becomes clothed in mathematical apparel. Insofar as the intuitive mind is the sole cause of its progress, it does not truly merit the name of science. Let us hear what Kant has to say on this score:

I propose that, in every particular theory of nature, there is nothing *properly* scientific except the amount of *mathematics* which it contains. After what has preceded, in fact, all science, properly speaking and above all the science of nature, requires a pure part which serves as the foundation for its empirical part. This pure part rests on knowledge *a priori* of natural things. Now, to know *a thing a priori*, is to know it in terms of its simple possibility. In order to know the possibility of natural things, that is, to know them *a priori*, it would be necessary further that *the intuition a priori* correspond to the concept that may be given, that is to say, the concept that we may construct. Now, rational knowledge through the construction of concepts is mathematical knowledge. Thus a pure philosophy of nature absolutely speaking, that is to say, one that searches solely for that which constitutes the concept of nature in general, could arrive at possible truth without mathematics. But a pure theory of nature, concerned with definite natural objects (theory of bodies and theory of the soul), is only possible in a mathematical mode. And since, in every theory of nature there is no true science except insofar as it contains knowledge *a priori*, the theory of nature will only then contain science properly speaking to the extent that mathematics shall be applied to it.

To the extent that a concept capable of being constructed (a *desideratum* which it is difficult even to gratify) shall not have been found for the chemical action of matter, chemistry cannot be anything but a systematic art or an experimental doctrine, but in no case a science properly speaking, for the principles of chemistry [in that case] are purely empirical and do not allow of being represented *a priori* in intuition. They do not in the least make the possibility of fundamental laws of chemical phenomena conceivable for they are not capable of being worked on by mathematics. . . .

The theory of bodies can only become a science of nature when mathematics is applied to it.[4]

In order that the German mind, then, should feel itself capable of scientifically treating a field of our experimental knowledge, it is necessary that deduction and, if possible, mathematical reasoning should be used therein. One understands, consequently, why the German has been the last of all in the vast concourse of civilized nations to enter into the establishment of a progressively perfect physics. One understands also how, in this concourse, this late-comer has taken such a prominent place. This is not fortuitous, but necessarily follows from the characteristics of the German genius, barren of intuition and powerfully mathematical.

Let us take, for example, a science whose rise in Germany over the last fifty years has been prodigious, namely, chemistry.

"Chemistry is a French science. It was established by Lavoisier, of immortal memory." Thus did Adolphe Wurtz express himself in the first lines of one of the finest discourses that could have been composed on the progress of science.[5] The studies by which Lavoisier set the basic foundations of chemistry are such that in them the intuitive mind expressed the full scope of its power and ability.

Chemistry developed for a long time before German laboratories contributed any outstanding discoveries to its progress. Their role is still slight in those researches which, through the instrumentality of the theory of types, go to make up our modern organic chemistry. Of these researches the brilliant pioneer is our J. B. Dumas. Immediately after him comes Laurent, who one day was to become the *premier doyen* of the Faculty of Sciences at Bordeaux; the Strassburger Gerhardt; the Englishman Williamson; and finally Adolphe Wurtz, another Strassburger, in whom intuition, vivacity, and the ardor of French genius all attained their acme. Among the creators of this science a single German, Hoffmann, the rival of Wurtz in the discovery of amines, can lay claim to a place of any significance.

But it happens that these admirable researches introduce to chemistry the language and processes of mathematics. In order to depict the constitution of organic bodies (in which carbon plays an essential role), the syntheses by which they can be produced, the substitutions which transform them into each other, the isomerisms and polymerisms which diversify them without changing their composition, the new theory bonds assemblages of points each to the other according to fixed rules. It was a German, Kékulé, who formulated these rules in a precise and systematic manner. The processes of reasoning by which the new chemistry explains or foresees reactions belong to a particular branch of geometry, the *analysis situs* (topology). The intuitive mind henceforth has a much diminished role in the development of organic chemistry. The assistance of the mathematical mind, however, becomes daily more indispensable. What Kant regarded as a "desideratum which it is difficult ever to satisfy", has come to pass. "The principles of chemistry . . . have become susceptible of mathematical treatment". At this point the German immediately takes possession of this part of chemistry, to which he devotes himself passionately. Thousands upon thousands of new organic compounds issue from his huge and manifold laboratories. In order to classify and describe these compounds, the German gives expression to principles drawn from topology by means of which he writes up treatises on chemistry which are exactly like works of mathematics.

The history of chemical mechanics allows of similar observations. The object of this science is to determine the influence which physical surroundings such as pressure, temperature, or the degree of concentration of solutions, exercise on the faster or slower rate, cessation, and change of direction of chemical reactions. The master-thought around which the fundamental hypotheses of this doctrine are grouped is this: the physical surroundings [*circonstances physiques*] which we have enumerated behave, as regards chemical reactions, as they behave as regards changes in physical state, such as the vaporization of liquids, the fusion of solids, the dissolution of salts and gases. To perceive this analogy, hidden under the many and obvious differences which separate chemical transformations from physical transformations; to display it through ingenious and convincing experiments; to warrant it against the objections its extreme novelty aroused; finally, to show its fecundity, required much force of intellect and subtlety of mind. After having instigated the divinations of Georges Aimé, this force of intellect and subtlety of mind directed the imposing work of Henri Sainte-Claire Deville and his disciples, Henri Debray, Troost, Hautefeuille, Gernez, and all these masters of the art of imagining and interpreting an experiment whose names I salute at this point with the heartfelt veneration of a grateful disciple.

After the school of Henri Sainte-Claire Deville had laid the experimental foundations of chemical mechanics, others came along who pushed this science to take mathematical form. In this regard it was sufficient for them to have recourse to the principles of thermodynamics, which already had brought changes of physical state, vaporization, fusion, and dissolution, under its laws. The first person who had the idea of applying the theorems of thermodynamics to chemical dissolution was my master, J. Moutier, soon followed by another Frenchman, Peslin. Shortly thereafter two men took the path which Moutier had opened: the German, Horstmann, and the American J. Willard Gibbs. But the latter pushed on much farther than the former, and the mathematical theory of chemical equilibria left his hands almost complete.

The question then was one of submitting the corollaries of this mathematical theory to the control of experiment in order to determine its exactitude and fecundity. This, in France, was the task of M. H. Le Chatelier. It was, however, principally the work of the Dutch school of M. Van der Waals, Bakhuis Roozboom, and Van't Hoff—Van't Hoff, who had been a student of Wurtz, Van't Hoff, of whom one of his Dutch disciples, M. Ch. Van Deventer, wrote:

> In many respects, Van't Hoff's work ought to be regarded
> as French rather than German. To be sure, he experienced
> a profound veneration for what is solid; but what he

passionately loved was the idea, the idea sketched out
with broad strokes, and his researches tended to hurl
ideas against the world, much rather than to take an enormous and massive block, a block that no one could move,
in order to round it off and polish it in every way. That
work he gladly left to others.[6]

These others set to work; they were Germans.

Now, in fact, chemical mechanics had reached the point at
which the mathematical mind could press its principles so that,
in a regular fashion, they might yield all their entailed consequences, and that these consequences should be, each in its
turn, following procedures henceforth fixed, submitted to
experimental testing. This systematic work—now that it was no
longer a question of invention, but of the orderly development
of an already imagined theory—the German laboratories set
themselves the task of accomplishing.

The history of stereochemistry would offer us a third
example of the truth that we have done our best to establish.

At the beginning of this history we would find the memorable
researches of Pasteur on tartaric acid and the tartrates. We
would see these researches establish a set correlation between
the crystalline form of a substance and the rotation which this
form imposes on polarized light. Next we would see the Frenchman
Le Bel and the Dutchman Van't Hoff simultaneously conceive
the bold thought that it was possible to transfer completely to
chemical formulas what Pasteur had shown to be the case for
the crystalline form. The observations of these two chemists
brought the first confirmations of their hypotheses. Since then,
the chemistry of substances endowed with the power of rotation would be governed by mathematical rules of extreme precision. It was at this point that the German laboratories would
take over the study of these substances. At this moment, with
the aid of the mathematical laws formulated by stereochemistry,
Emil Fischer and his students would establish the chemical
composition of sugars and bring about their synthesis.

When an experimental science has come to the point of perfection at which deductive reasoning unrolls at length the consequences of hypotheses, or, better yet, when it lends itself to the use of mathematical reasoning, it allows man to foresee perfectly exactly what will be produced in given circumstances, and its expectations are almost assured against failure. Now, foresight is power. Consequently, experimental science, when it has become deductive, above all when it has become mathematical, is the guide of industry.

From this point, in order to promote industry, the mathematical mind must draw out all the corollaries entailed in the principles of science so that the engineer may find among the corollaries a mass of useful truths, practical recipes, and patentable processes. Is it not clear, consequently, that the mathematical mind of the Germans, so fitted to deduce all the consequences from a given principle, was marvelously adapted to extract an industry of extraordinary power from our mechanics, our physics, and our chemistry—the moment they reached a deductive and mathematical stage? It has often been noted that the German, who is little capable of new ideas, was the most skillful at bringing together and developing the consequences of inventions which came from elsewhere. These are really, in fact, the characteristics by which the excessive development of reason or the mathematical mind [*raison ou l'esprit de géométrie*] has repressed common sense and has never permitted it to grow into the intuitive mind.

When a reason such as this sets to work on the progress of the experimental sciences, it finds itself greatly exposed to two failings.

The first is to impose the deductive form, or even the form of mathematics, on a science of observation which is not yet ready to take on that form. From hypotheses which the intuitive mind has not yet taken the time to analyze and make precise, which experiment has not controlled often enough to assure their firmness, numerous and detailed consequences are hastily drawn by quite rigorous processes of reasoning. This amounts to erecting a substantial building on quicksand. It is to do a vain thing and to head for disaster.

The second of these failings is to forget that a science whose principles have been drawn from experiment still remains adjudicable by experiment. When, then, it comes to taking some of the consequences deduced from theory and comparing them with the facts, the observer ought to inquire into reality with as much care and impartiality as if the process of reasoning involved had not offered him any preliminary indication. His attention ought to be brought to bear with a particular sharpness on every fact which, insignificant as it might appear, deviates from what has been predicted. His intuitive mind ought to collect and weigh these evidences with scrupulous precision, always ready, if these contradictions of what was expected should demand it, to condemn the theory, despite all the confirmation it may previously have received. To believe that rigorous deduction is capable of conferring on the corollaries of a hypothesis an *a priori* certitude which the premisses did not enjoy; strong in this confidence and without further information, to hold as accidental and negligible all discrepancies between the predictions of the theory and the reality; or even further, to think up for these discrepancies explanations and excuses forged solely for partisan purposes, is to fall into the most serious of blunders, if not into the most culpable dishonesty.

A reason which, for lack of good sense and an intuitive mind, fails to recognize the point at which the truth bursts forth; which so willingly sees in the consequences of the rigorous deduction at which that reason excels the source of a certitude which the premisses do not enjoy—such a reason, as we say, is quite ill-protected against the two dangers which we have noted.

Have some new phenomena been discovered in the domain of physics? This sort of reason does not wait until such time as repeated ingenious experiments, sharply analyzed, severely criticized, may have established, clarified, and made more precise the laws pertaining to the phenomena. To accomplish this task properly would require too much subtlety of mind [*finesse d'esprit*]. The reason in question replaces these facts which have only just been observed, with algebraic equations, and deduces endlessly from principles of uncertain worth. How many theories of this sort, devoted, for the most part, to some electrical effect, have been brought to us from Germany over the past twenty years!

And on the other hand, in each of these laboratories, as huge as factories, there work a constellation of students [*une pléiade d'étudiants*] with a military discipline. Each of them aspires to acquire the envied title of 'Doctor' in a reasonable time. Each candidate receives one of the numerous but similar inferences from a theory. The testing of each of these inferences will afford the matter of the students' inaugural dissertations, the slender thesis which their doctorate will crown. The theory is always verified, without complications, without incident, in the allotted time.

In French laboratories theories do not always display such docile complaisance. Complete as they may be, and well tested by previous experiment, they endlessly reveal themselves as over-simple. Reality is so ample and so complex that the theories are outflanked by it in all respects. Never can the sagacious observer pursue the testing of a theory for a considerable period of time without discovering unforeseen, difficult, or exceptional cases, upon which his subtlety of spirit finds many occasions to exercise itself. By contrast, at the threshold of certain German laboratories one could write, as on certain lotteries in fairs: Here, everyone wins. Seeing at each throw that the dice bring a double-six, will you exclaim, with Pascal, that "the dice are loaded"? Do you think you are carrying on business with card-sharpers? No, you have well-disciplined mathematicians before you. When a theory is accepted by *Herr Professor* and is thus true, they cannot conceive how the consequences which can be rigorously drawn from it could be false.

Let us give a particular example of the two shortcomings with which we have reproached German science. Haeckel's attitude to the Darwinian hypothesis furnishes such an example.

Charles Darwin was a marvellous observer. Patient and sagacious, he studied with as much thoroughness as penetration certain variations found in individuals of the same animal or vegetable species. It appeared to him that within such a species, the formation of various races could be explained by the play of natural selection. Amplifying, then, but prudently, the law which he had thus discovered, he thought that it made conceivable the gradual derivation of living beings of all species from a single source [*souche*].

This hypothetical proposition appeared to make sense of a great number of facts. Did one come across any facts in nature with which it was incompatible? If so, it was the task of observers to gather them together with great care, to examine them closely, and to determine whether their evidence against the theory of natural selection was merely apparent, or whether it was in formal contradiction to it. This is a labor which many naturalists carried out, in the first rank of which stands a master whose perspicacity Charles Darwin praised highly, our great Henri Fabre. Of the Darwinian hypothesis, this work left hardly anything but debris.

Haeckel did not take the question from the same point of view as the observers. Here is what he wrote on the subject of the Darwinian theory thirteen years after the appearance of the *Origin of Species:*

> It is not up to the fancy of each zoologist or botanist to
> accept or not accept it as an explicative theory. One is
> rigorously obliged, by virtue of the fundamental principles
> in force in the domain of the natural sciences, to accept
> and conserve, insofar as a better one is not offered, every
> theory, which is capable of being reconciled with efficient
> causality, even though it be weakly established. Not to do
> so is to reject every *scientific explanation* of the phenomena.[7]

The unique title which a scientific hypothesis could have to our belief depended, we thought, on the agreement of its consequences with all facts carefully observed. Not at all! "Weakly established" as it may be, it "rigorously obliges" us to accept it, at least if we are not in possession of a more satisfactory hypothesis. That, as you say, is not in accord with common sense. Have I not said to you that, in the reasoning processes of many Germans, good sense is too often missing?

The axiom of natural selection being thus posed, the mathematical mind then reels off the corollaries, and, for better or worse, nature is thereby obliged to agree with the consequences regularly deduced from a principle which we are "rigorously obliged" to accept. How marvellously relaxed is Haeckel's attitude to experience!

Let us read, for example, the lecture which he devoted to spontaneous generation. The purpose of this lecture is to prove that spontaneous generation is possible. Hypotheses which concern the formation of the system of the world, considerations relating to the antiquity and the temperature of the Earth, relations between organic and inorganic chemistry, a description of the legendary *Bathybius haeckelii*, are successively invoked in support of the thesis which is being defended. Nevertheless, we remember that experimenters believed they had proved not just the mere possibility but the actual reality of spontaneous generation. We recall that Pasteur, in a memorable series of researches, convicted such experimenters of their error. One must certainly take into account this celebrated debate. Haeckel wrote 29 pages on spontaneous generation without mentioning the name of Louis Pasteur. Concerning this dispute, here is all that Haeckel wrote:

> Up until now, neither the phenomenon of autogenesis[8] nor that of plasmogenesis[9] have been directly and incontestably observed. Then and now numerous and often quite interesting experiments were devised in order to verify the possibility, the reality, of spontaneous generation. But these

experiments, in general, dealt with plasmogenesis rather than with autogenesis, with the spontaneous formation of an organism at the expense of organic materials. It is evident that, for our history of the creation, this last category of experiments is only of secondary interest. Does autogenesis exist? That, above all, is the question which it is important to resolve. Is it possible that an organism could be born spontaneously from matter which was not previously alive, from strictly inorganic matter? We can now neglect all the quite numerous experiments attempted with such ardor during the last ten years concerning the subject of plasmogenesis, and which, besides, have had *for the most part*[10] a negative result. If, in fact, the reality of plasmogenesis were rigorously established, that would prove nothing concerning autogenesis.[11]

Indeed. But that it should have been impossible to produce spontaneous generation, even out of substances furnished by living beings, gives a singularly strong presumption that it will be impossible to obtain it from purely mineral substances. An honest man would have told us so.

When a scientist has recourse to conjuring tricks in order to make an experiment which inconveniences him disappear, it is not good sense that he lacks; it is good faith.[12]

LECTURE III:
THE HISTORICAL SCIENCES

LADIES AND GENTLEMEN:

Historical truth is a truth of experiment [*vérité d'expérience*]. In order to recognize, or to disclose an historical truth, the mind follows exactly the same path as it does to disclose experimental truth. Rather than observing facts, however, history studies monuments, it deciphers texts. Moreover, these monuments and these texts are themselves also facts.

At the beginning of all historical research, as at the beginning of all experimental research, a preconceived idea is necessary. This idea has often been suggested to the historian through some lucky find—the discovery, for instance, of some monument until now buried in the ground, or of some unknown text, which chance brings him to dig up in the debris of an ancient city or the dust of a library.

It is necessary to submit this preconceived idea to the control of the documents and, in order to do that, one must research these documents. Such research is often difficult, always entrancing, but has no precise rule to direct it. In it one rediscovers the fascination and the unpredictability of the chase. Precisely where everything appears to promise a rich quarry one finds the thicket empty, and the game takes off from its

41

cover where one never thought to come across it. Directing
such a chase smacks so little of the reasoning of reason that it
is tempting to compare the skillful searcher of excavations and
archives to a dog who follows a scent, and to say of such a
person that he has a flair for such things.

One must make use of the documents collected. Each of
them necessitates, then, a discerning scrutiny. Is it authentic?
Was the date it bears, the signature it exhibits, not added after
the fact by some forger or some ignoramus? Is it complete? Or
rather, isn't it only a fragment and, in that case, what might be
the extent, the nature, the meaning of the missing parts? Is it
candid? Has the author said, without addition and without
reserve, all that he thought to be true? Did his passions and his
interests not lead him to exaggerate or conceal or modify part
of the events which he retold? Was the author well-informed,
or rather, on the contrary, was he in no position to know
thoroughly those things which interest us the most? Do we
understand with exactitude the language he uses? Do the
thoughts which he sets forth convey to us adequately the sense
they had for those to whom he addressed them? Such are the
manifold problems, here touched upon only in passing, which
the slightest of documents present, problems which it is
necessary to resolve if one would transform this dead thing,
these several marks engraved on stone or metal, on papyrus,
parchment or paper, into a living, speaking being which tells us
about ages past. It is necessary to know how to solicit a text,
said Fustel de Coulanges. The adversaries of this master have
pretended to see in this precept the advice of a suborner of
witnesses. As if he had been capable of such an improper
thought, the man whose probity extended to ingenuousness,
the historian who made this memorable remark about the
science to which he had vowed his life: "We demand of it the
spell of perfect impartiality, which is the chastity of history!"[1]
Let us listen more attentively to the words of Fustel de Coulanges.
In the presence of a text, the historian ought to be like the

examining magistrate confronted by a witness who saw things inaccurately, or who obstinately refuses to recount what he saw, or who chooses to invent things which he did not see. The magistrate, however, by virtue of prudent, patient, skillfully-connected questions, ends by drawing out of this ignorant or recalcitrant or dissembling witness precise, truthful, and useful information.

When one has forced the texts to speak, it is then necessary to listen to their language. Their deposition does not present only those things which are favorable to the preconceived idea concerning which they were called as witnesses. Within it are items which will tend to weaken our preconception. Should the testimony of those items outweigh favorable testimony? Is it necessary then to condemn and reject immediately the preconception in which our mind had thought it caught sight of a glimmer of truth? The job in hand now is that of a magistrate. It calls for all the qualities which make the just magistrate, not only mental rigor and insight, but beyond these that fine and rare virtue of the heart which is called impartiality. This impartiality is often quite difficult to practice. It is hard to give up the idea to which we have initially been predisposed, because man is always attached to his own opinions. It is hard often and most of all because the historical proposition which we wished to establish was of use in defense or attack, in defense of a cause which is dear to us, or to attack a doctrine which we find detestable. In the realm of every science, but more particularly in the realm of history, the pursuit of the truth not only requires intellectual abilities, but also calls for moral qualities: rectitude, probity, detachment from all interest and all passions.

Once our initial supposition has been rejected, we must make another one, which takes account of all the texts and all the monuments already known. Then, if possible, we must check this second proposition against new documents. In such fashion, by the continual comparison of our thought with facts, by this incessant impression of facts upon our thought, little by little, a historical truth will be found to disengage itself, to become explicit, to become clear. In order to justify a hypothesis concerning the origins of the Carolingian monarchy, a historian need proceed in no other fashion than in that which Pasteur used to verify a hypothesis concerning the cause of rabies.

Historical work essentially requires the intuitive mind for its accomplishment. It is appropriate that one can say of such research that its "principles are found in common use and are open to the scrutiny of everybody. One has only to look, and no effort is necessary; it is only a question of good eyesight, but the eyesight must indeed be good, for the principles are so subtle and so numerous, that it is almost impossible but that some escape notice. Now, the omission of one principle leads to error; thus one must have very clear sight to see all the principles, and then an accurate mind not to draw false deductions from known principles."[2]

Deprived of the spirit of finesse, the German intellect is singularly myopic. Nevertheless, it has desired to apply itself to historical work, to become master of it, and to teach others how it ought to be pursued. Thus, it has claimed to trace out a path for the historian's work so narrow, so rigorously bordered by guard-rails that it could be followed blindly. It took it into its head to reduce documentary research, textual criticism, and proving conclusions to rules so precise and so peremptory that intellects most lacking in subtlety and most deprived of common sense had only to follow them in order to arrive unerringly at truth. Thus the hands of a watch, which do not perceive the time, are constrained by the precisely stamped and meshed mechanism to mark the time exactly. The entirety of these rules which transform the critique of the historian into well-regulated mechanically functioning clockwork has been offered to the admiration of the world under the name of the historical method.

There is not, there cannot be, any historical method.

Whoever says *method* says 'a manner of procedure traced with precision, which is capable of leading without deviation from one limit to another'. In the arts, method exists wherever there is an explicitly formulated procedure which, with the help of specific means, allows one to execute unerringly a prescribed work. In mental operations, method exists if reason possesses a rule of conduct which leads it faultlessly from the knowledge of certain given truths to the discovery of other truths which are necessary consequences of those given truths. Now, "a process of reasoning in which, certain things being given, something else necessarily follows by reason of the nature alone of the given things themselves" is, by the very definition of Aristotle,[3] a syllogism: "*Syllogismos de esti logos en hō tethenton tinōn heteron ti tōn keimenōn ex anagkēs symbainei tō tauta einai.*" This is as much as to say that, in the realm of the intellect, 'method' is synonymous with reasoning syllogistically and that it is exclusively deductive.

Method takes possession of a science, then, precisely at the moment that this science comes under the sway of the mathematical mind. Insofar as the progress of a science depends on subtlety of mind alone, that science is in revolt against all method.

There is a general method of deduction, of which Aristotle permanently established the laws. There is a method particular to each of the sciences which deduction develops: a method for Algebra, one for Geometry, one for Mechanics and Mathematical Physics. From the day the atomic notation allowed one to say, by means of precise rules, by what series of reactions one could unerringly bring about this substitution or that synthesis, there was a chemical method.

There will be no historical method at all as long as history does not proceed by deduction; and history will never be a deductive science because man, its subject, is too complex, altogether too difficult to pin down with any definition, moving as he does in a milieu of events too numerous, too fine, too confused to be measured.

The best-placed witness has not seen everything. Who will specify the concourse of trifling circumstances thanks to which a given fact escaped him? This general did not notice a certain episode of the battle for which he was responsible: he was busy controlling his horse, which had just been bitten by a fly.

The most veracious witness does not report everything he has seen; he relates only that which appears to him to be worth noting. And how petty the motives for his preferences often are! Of a feat of arms marking the end of a battle, the general made no mention whatever in his account. He had given full details of an action completely similar to the one in question which took place at the beginning of the battle. Why the difference? He cut short the end of his account—because he was ready to drop from fatigue.

Does one expect then to attain by rigorous reason to the exact connection between what happened and what a witness noticed, between what he saw and what he reported? Where in this process are those simple ideas, those clearly defined notions, to be found? Those few rudimentary principles without which the deductive method cannot be followed?

Another reason forbids history the use of the deductive method.

In order for a science to be able to become deductive it is necessary that, in the realm which it explores, the consequences must follow necessarily, *ex anagkēs*, from the data. It is necessary that this realm be governed by a rigorous determinism.

So one can never deduce in history; never affirm that such causes, which are known, have necessarily produced such a result. Always, indeed, the human will is inserted between causes and that which comes of them, and this will is free.

It is impossible, for example, to formulate an infallible procedure for recognizing whether the testimony given by a document is genuine or mendacious. Gather together all the reasons which pressed the author to dissimulate the truth; cite all the interests which solicited him, all the passions which agitated him, all the vices which corrupted him: he yet remained free not to mislead, and therefore it was possible that he was speaking the truth. If you accuse him of duping us, it will be because your common sense, your subtlety of intellect, make you suspect him of false witness. This will not be the conclusion of a syllogism, for the free will of this man would always prevent your syllogism from holding.

For historical criticism to function with the same surety and precision as a well-regulated mechanism, it would be necessary that man himself be a machine, that he have the simple and rigid wheelwork of a machine, and its necessary movement.

Now German history aims at being methodical history, deductive history. From principles which it sets forth as certain, it claims rigorously to draw consequences that cannot fail to be true, to conform to reality. And if the facts do not agree with the corollaries of reason, so much the worse for the facts. It is they that are in error, not the syllogistic conclusions. It is they that will be retouched and corrected, not the predictions which the method supplied.

Alas, it would be impossible to maintain that French historians have never displayed this bad habit. The truth, in many circumstances, can serve or obstruct interests which are so strong and passions so violent that it is difficult to search for it in complete independence, difficult never to be tempted to model reality after the image fashioned by the thesis which one means to sustain, rather than to shape the thesis upon facts. "For fifty years now our historians have been party men," Fustel de Coulanges wrote in 1872.[4] "As sincere as they may have been, as impartial as they may have believed themselves to be, they followed one or another of the sets of political opinions which divide us. Passionate searchers, powerful thinkers, skillful writers, they put their zeal and their talent to the service of a cause. Our *history* resembles our legislative assemblies: one can make out right, left, and center parties. It has been a tournament-yard where opinions did battle. To write the history of France was a means of working for a party and combating an adversary. History thus became among us a sort of permanent civil war."

To this combative history, such as it was at the time he wrote, Fustel set in opposition historical science such as his uprightness wished it to be. "It would be preferable," he said, "that history always have a more irenic manner, that it remain a pure and absolutely disinterested science. We should wish to see it soar in that serene region where there are neither passions nor rancor nor the desire for vengeance. We ask of it the charm of perfect impartiality, which is the chastity of history . . . The history which we love is that true French science of other days, that erudition so calm, so simple, so exalted, of our Benedictines, of our Academy of Inscriptions, of Beaufort, Freret, and many others, famous or anonymous, who taught Europe what historical science is, and who sowed the seeds, so to speak, of all present-day erudition. History, in that time, knew neither partisan nor racial hatred. It sought only for truth, praised only the beautiful, detested only war and covetousness. It served no cause, it had no fatherland. Not teaching invasion, it had no need to teach revenge."[5]

The aspiration that Fustel de Coulanges formulated has been heard and answered. This was so in the first place because the master, in his teaching as in his writings, gave a perfect example of the virtues to which he laid claim for the historian. In the second place it was so because he addressed his hope to the French: and if we are too given to fighting among ourselves, there are at least many of us who would be ashamed to use unfair tactics. Also we have seen born and mature a constellation [*pleïade*] of historians who have renewed the tradition of pure and calm impartiality saluted by Fustel in the French creators of history.

In order to be genuine, they had to know how to set aside political passions, even if their impassive attitude raised the protestations of all parties. Surely many among us can recollect the contradictory movements in public opinion aroused by publication of the successive volumes of the *Origines de la France contemporaine*. With his first volume, *L'Ancien régime*, Taine deeply irritated the royalists. The three following volumes, devoted to the history of *La Révolution*, provoked a tempest not yet allayed among the Jacobins. Finally, the appearance of the first volume of *Le régime moderne* was greeted by the imprecations of the imperialists. In this work, no one found the flattering portrait of his own party which alone would have been judged a true likeness. Hippolyte Taine, great and honest man that he was, did not desire to sketch out such an accommodating image. "In my judgement," he wrote, "the past has its own contours, and the portrait which is presented here only resembles that of France in former days. I have traced it out without having present-day debates in mind. I wrote as if I had as a subject the revolutions of Florence or Athens. This is history, nothing more, and truth to tell, I thought too highly of my trade as a historian to tell another dissembling tale alongside it."[6]

Without ceasing to be men, and men of spirit, without severing themselves from the passionate devotion which each of them bore for causes which they held to be just and good, numerous historians such as Fustel de Coulanges or Taine have known how to retain an unflinching respect for truth. One Henry Houssaye, for example, could write at the beginning of his admirable *1814*:

> We have conscientiously sought the truth. At the risk of rumpling everyone's opinions, we have wished to omit nothing, to hide nothing, to soften nothing. But impartiality is not indifference. In this account, in which we have above all focussed on France, the great casualty, we have not been able to avoid quivering with pity and anger. Without siding with the Empire, we have rejoiced at the victories of the Emperor and suffered through his defeats. In 1814, Napoleon is no longer the sovereign: he is the general. He is the first, the greatest, the most resolute of the soldiers of France. We rallied to his colors, saying, with the old peasant of Godefroy Cavaignac: "It is not a question of Bonaparte any more. The soil [of France] has been invaded. We go to fight".[7]

And when French soil was invaded once again, Houssaye himself was to fight.

The subtlety of mind of a Henry Houssaye knew how to harmonize love of country pressed to the point of sacrifice with love of truth pressed to the most watchful impartiality. Do not ask such a masterpiece of tact and delicacy of the mathematical mind of a German.

The mathematical mind has transformed history into a deductive science. It emanates from axioms which this mind holds to be absolutely true. The corollaries which it rigorously deduces from these axioms cannot fail themselves to be absolutely true. It is therefore certain in advance that all the documents gathered with so much patience and attention to detail have to be arranged within these limits constructed *a priori*. The German historian remains so assured by this necessity that if, by some chance, some text does not fit completely into the spot allocated for it—well, a brutal shove will make this rebel conform to the discipline which reason imposes on it.

Among the axioms to which German historiography [*l'histoire germanique*] appeals there is one which dominates all the others. It is set forth in these words: *Deutschland über alles!* Germany over all!

It is not just recently that German historians have reposed an absolute confidence in this principle. Listen to what Fustel de Coulanges wrote of them on the morrow of the war of 1870–71:

> A unique and common will animates [*circule*] this great
> learned body, which has only one life and one soul.
>
> If you look for the principle which gives this unity and
> this life to German erudition, you will note that it is love
> of Germany. We claim in France that scholarship knows
> no fatherland. The Germans frankly maintain the opposite
> thesis. "It is false," one of their historians, M. de Giesebrecht,
> wrote recently, "that science knows no fatherland, that it
> soars above frontiers. Science ought not to be cosmopolitan.
> It should be national, it should be German. . . . ''

The German scholar has a passion for research, a capacity for work which astonishes us French. But don't believe that all that ardor and all that work are for science. Science here is not the end, it is the means. Beyond science, the German sees the fatherland. These scholars are scholars because they are patriots. The interest of Germany is the final goal of these indefatigable investigators. One could not say that the true scientific spirit is absent from Germany, but it is much more rare than one would have thought generally. Pure and disinterested science is an exception there and is only moderately appreciated. The German is a practical man in all things. He wants his erudition to serve some cause, to have a goal, to hit home. It must more or less march in concert with national ambitions, with the desires and aversions of the German people. If the German people desire Alsace and Lorraine, it is necessary that German science, twenty years in advance, take possession of these two provinces. Before Holland is seized, history must show that the Dutch are Germans. It will show as well that Lombardy, as its name indicates, is a German land, and that Rome is the natural capital of the German Empire.

What is more singular here is that these scholars are perfectly sincere. To impute the least bad faith to them would be to calumniate them. We do not think that there is a single one of them who would agree knowingly to write a falsehood. They have the best will to be veracious, and they make serious efforts to be so. They surround themselves with all the precautions of historical criticism in order to force themselves to be impartial. They would be so, if they were not Germans. They are not able to make their patriotism other than the most forceful. It is said with some reason beyond the Rhine that the conception of truth is always subjective. The intellect only sees, indeed, what it wants to see. The eyes of German historians are made in such a fashion that they perceive only those things which are favorable to the interests of their country. Such is their fashion of understanding history. They could not comprehend it otherwise. Hence in their hands the

history of Germany became, quite naturally, a veritable
panegyric. Never was a nation so highly vaunted. They
quite skillfully profited by the reproach of boastfulness
which we brought against ourselves, in order to boast
about themselves with impunity. We proclaimed ourselves
braggarts; they boasted openly. We made the whole world
aware that we were boasting, even when our historians
appeared to work hard at disparaging us. They boasted
without warning anyone, modestly, humbly, scientifically,
despite themselves and out of sheer necessity. This has
gone on for fifty years.[8]

Forty-three years have passed since Fustel wrote that. The
Germans have continued to boast about themselves. But their
tone has changed from one of humility and modesty to become
one characterized by arrogance, presumptuousness, inordinate-
ness. They used to murmur sweetly: *Deutschland über alles!* The
proclamation of their favorite axiom has now become the furious
howling of a pack of mad wolves.

Documents patiently gathered and scrupulously criticized
by German scholarship should all bear out the establishment of
this proposition: Everything that is great, beautiful, or good in
the world is German. Just how to set about constraining the
texts to bear the witness expected could be shown you by a
professional historian with more examples and more compe-
tence than I can. Limited as my excursions in the realm of
history have been, however, they have exposed me to several
curious examples of this art of adapting texts. Allow me to give
you one. It is from Dr. Jos. Ant. Endres's book on Honorius
Augustodunensis.[9]

There exists a work by a twelfth-century writer entitled *De
Luminaribus Ecclesiae*. In it the presiding geniuses who have been
the torchlights of the Church are successively enumerated, along
with their own principal writings. Its last chapter is devoted to
the author himself. He tells us that he is *Honorius, presbyter et
scholasticus ecclesiae Augustodunensis*. In the judgment of everyone,

including M. Endres himself,[10] no town in Germany was ever called Augustodunum. The only town which, historically, has borne that name is the French town, Autun. There is therefore no doubt about the meaning of the Latin description. It ought to be translated: 'Honorius or Honoré, priest and school-inspector [*ecolatre*] of the Church of Autun'.

If I now were to ask you to what region did Honorius belong, you would surely respond to me, 'Autun, of course!' You understand nothing of the historical method. M. Endres, who knows it, responds without hesitation: Honorius of Autun was from Ratisbon.

No doubt you wish to know how such a conclusion is established. You shall see.

The axiom which supports the whole demonstration is set forth in the first line of M. Endres's work: Honorius of Autun was a German. Why? The German authors, Rupert von Deutz, Gerhoh von Reichensberg, Otto von Freising affirmed, without any further proof, that he ought to be counted among the number of the most celebrated German writers. M. Endres agrees with their judgment, with no trace of discussion.

Since Honorius of Autun was a German, it only remains to determine the town in Germany which gave him birth.

In the very article wherein he calls himself a priest and schoolmaster of the church of Autun, Honorius claims to be the author of *The Image of the World,* a summary and elementary description of the universe which had an extreme vogue in the Middle Ages. *The Image of the World* contains a brief précis of geography. Let us look at it. We see that only one single town in Bavaria is noted. This is the town of Ratisbon. It is beyond doubt: Honorius of Autun was from Ratisbon. Doberentz proposed this conclusion, and M. Endres accepts it with full confidence.

Endres is not however able to conceal from himself[11] that this conclusion runs into a great difficulty. In a text which M. Endres regards as authentic, one that he uses as the basis for his study, our *German* from Ratisbon is called priest and schoolmaster of the church at Autun. But for such a trifle we do not need to renounce the conclusion of our rigorous demonstration. We content ourselves with suggesting this explication: "Might we not be dealing with a sort of medieval pseudonym?" (*War es nicht denkbar . . . dass wir es also mit einer Art mittelalterlichen Pseudonymie zu tun haben?*) Thus, a note tells us Honorius Conrad d'Hirschau, a contemporary of this Honorius of Ratisbon who took a malign pleasure in calling himself "of Autun", signed his tracts, "The Pilgrim", *Peregrinus*. Ah! But Herr Doktor Endres, he did not for all that go so far as to call himself priest and schoolmaster of the church of Chartres, or of that of Carpentras.

We have here an example of the truly disconcerting method by which German historians, despite the most explicit and most clear texts, annex to Germany a man or a territory.

At times, moreover, they do not go to the trouble of finding similar subterfuges. When a text inconveniences them, they suppress it, purely and simply. In his fine discussion on true and false patriotism, given recently at Bordeaux under the auspices of the *Journal des Débats*, M. Camille Jullian cited a veritably "colossal" case of such an amputation of documents. In many passages, Julius Caesar's *Commentaries* clearly affirm that Gaul extended as far as the Rhine. Recent German editions of the *Commentaries* have suppressed these passages, purely and simply, as apocryphal. They couldn't be genuine, since a process of reasoning as rigorous as geometry shows that Alsace and Lorraine have always belonged to Germany.

Do you now perceive the reason for certain actions which revolt you? To the protestations which are raised on all sides against the atrocities committed by the German army, the German universities have, with one accord, opposed a manifesto whose process of reasoning can be summarized as follows:

We, the German Universities, are perfectly virtuous. Consequently, our teaching is perfectly virtuous.

Since that is the case, the Germans who have been formed by this teaching could not fail to be perfectly virtuous.

Therefore, they have not committed the horrors with which you reproach them.

And we, Frenchmen, reply: But look then at the smoking ruins of burned cities, the bodies of murdered women and children! It is a waste of time. They simply will not hear us. They are certain that their syllogism is conclusive.

If you should say to me that you have a right-angled triangle in which the square of the hypotenuse is not equal to the sum of the squares of the other sides, I wouldn't listen to you. You might well cry out to me, 'But look at it!' I would only turn away. Geometry, more certain than your sense perception or mine, would assure me that you were mistaken. To your protests I would be as deaf as the Germans are to the protestations of universal conscience.

Here we come, I believe, to the foundation of the German intelligence.

In any soundly constituted process of reasoning, "principles are intuited; propositions are inferred" [*les principes se sentent; les propositions se concluent*].[12] The axioms condense in themselves everything that common sense, sharpened by subtlety of intellect, has been able to discover concerning truth. Without adding the least parcel of truth to this storehouse, deductive reasoning faithfully distributes to the conclusions the riches which it has borrowed from the axioms.

The German turns all this around, for he is demented. His reason is a monstrous thing, wherein an excessive development of one faculty has aborted the other. Endowed with a powerful geometrical intellect which allows him to deduce with extreme rigor, he is deprived of common sense, of that subtlety of intellect which supplies the intuitive knowledge of truth. Therefore, he reverses the normal conditions of human knowledge. Incapable of judging whether a principle is true or false, he holds every axiom as a postulate, that is, as an arbitrary decree postulated by our will. Then, confounding truth with rigor, he holds as true every consequence deduced by rule from such premisses.

This is as much as to say that he holds as assuredly true every judgment the truth of which accords with his interests or his passions. In fact, it suffices that he propose to himself an axiom such that a formally rigorous deduction can draw from it the desired proposition.

For example: in times of war, the fancy enters his mind to massacre inoffensive beings? He sets forth this postulate: Everything that tends to shorten the duration of war is humane. Then, after having unrolled several quite conclusive syllogisms, he robs, violates, pillages, burns, executes, and torpedoes with the serene conscience of a benefactor of humanity.

With a marvellous perspicacity, Fustel de Coulanges recognized this: "The German is a practical man in all things. He wants his erudition to serve some cause, have some goal, strike home somewhere." German science, and above all history, is, quite often, only an arsenal from which the German supplies himself with principles appropriate to a justification of his actions. Thanks to the immense labor of his scholars, his philosophers, and his historians, the German has always at hand, at the moment of committing a crime, the axiom from which a solid process of reasoning will demonstrate to him that he is acting properly. Among scoundrels, this is most dangerous. Against remorse they have laid hold of an assurance as certain as that two and two make four.

Lecture IV:
ORDER AND CLARITY.
CONCLUSION

The German mind is powerfully mathematical, but it is only mathematical. This formulation sums up all we have said concerning the characteristics of the sciences of reasoning, the experimental sciences, and the historical sciences in Germany.

If the hallmark of German intelligence is this exclusively mathematical mind, what then is the source of that chaotic confusion, that profound obscurity, which the works produced by that intelligence so often display? For, finally, what is more clear, what more strictly arranged, than geometry?

Observe the concert-goer. His ear is quite subtle and experienced, and distinguishes with a marvellous exactness the slightest nuances of pitch or timbre. It resolves the most intricate chords. The harmony and the melody hold no secrets for it. The concert concluded, this man prepares to leave. He conducted himself skillfully through the complexities of the music. Now, he proceeds cautiously, hesitates, bumps into people and things. But don't be astonished. He is blind.

In an entirely similar fashion, German science is neither obscure nor confused concerning that which derives from the mathematical mind. German albegraists—a Weierstrass, a Schwartz—put into their research an admirable order. In it they push the concern for clarity to a fault. But when, leaving the proper domain of deductive method, German reason roves into the regions where an intuitive mind alone will be clear-sighted, it walks blindly.

Then, indeed, it imitates the blind. In circumstances in which the common run of men use sight to conduct themselves, the blind have recourse to the only senses they possess, hearing and touch. In this fashion, deprived of the light of common sense and that of an intuitive mind, German science tries, in areas where this light would be indispensable, to proceed according to the geometrical method. But this method cannot give it the light it needs.

Just as there are two sorts of minds, the intuitive mind and the mathematical mind, each of them contributing that which is peculiar to it to the construction of science, that without which the work of the other could never be completed; so too are there two kinds of orders, the mathematical order and the natural [or real: *naturel*] order. Each of these orders is a source of enlightenment when it is applied where it is appropriate. But one immediately falls into error if one imposes a natural order on materials which fall under the jurisdiction of the mathematical mind. And one would remain in profound darkness if one required the mathematical method to clarify that which pertains to the intuitive mind.

Following the mathematical method means never advancing any proposition which could not be demonstrated by means of propositions antecedently established.

Following the natural method means bringing together one with another truths which affect things analogous by nature, and separating judgments which concern dissimilar things.

In geometry itself it is sometimes necessary to take account of the natural method. Indeed, without in the least forfeiting the exactness that constitutes the entire method of geometry, it happens that it is possible, for the same set of theorems, to conceive several different arrangements. The intuitive mind, in this case, will suggest to the mathematician [*géométré*] which of these dispositions is the most natural, and therefore the best. This is an essential task, and, often, quite neglected by the mathematician [*géométré*] who is nothing but a mathematician. Inspired by Descartes and Pascal, *The Logic of Port Royal* already reproached such a mathematician. Among the faults it accused him of is this:

> *To have no concern for the true order of nature.* Here is to be found the greatest shortcoming of mathematicians [*géométrés*]. They imagined that there was virtually no other method to observe than that the first propositions should be able to serve to demonstrate those that follow. And so, without troubling about the rules of true method, which consists in always beginning from the simplest and most general things, so as to move next to the most complex and particular, they mix everything together and deal in a haphazard way with lines and surfaces, triangles and squares, proving the properties of single lines by figures, and making an infinite number of other inversions which disfigure this beautiful science. The *Elements* of Euclid are just full of this fault . . . [1]

It will not be astonishing that German mathematicians should have been much given to this failing. But for fear of being too technical, we might show here, by several examples, how the exclusive pursuit of algebraic precision [*rigueur*] often leads mathematicians [*géométrés*] on the other side of the Rhine into the most complete disdain for the order which, in the subject treated, would be imposed by natural affinities. But we would have to enter into details too numerous and too particular for our present purposes.[2]

When it is deprived of assistance from the intuitive mind and claims to be self-sufficient, the mathematical mind is not only incapable of arranging a mathematical theory according to the natural order, it also fails to recognize the affinity which exists between the various sciences. It overlooks the essential connections which link mathematics to the other parts of human knowledge.

The study of nature, astronomy, or physics necessarily poses problems for mathematicians which they will try to resolve. And the solution of these problems ought to be so directed as to serve the sciences of observation that call it forth. Pure mathematicians [*géométrés*] are often tempted to break this connection between their chosen science and the other sciences. Under the pretext that their meditation ought to be completely disinterested, they claim to set problems for themselves which they will subsequently resolve without the least intention of applying them to anything whatsoever, and "for the sole honor of the human mind". Nothing is more dangerous than acting in such a fashion. Not only does it deprive the sciences of observation of the indispensable means of research without which they would fall into pragmatic fact-collecting; but, further, in isolating the mathematical sciences it renders them sterile. Most questions which have proved to be fecund, which engendered broad theories of geometry or algebra, have been posed to the mathematician by the physicist or the astronomer. In a multitude of cases the science of observation was not content to formulate the problem, but also suggested its solution. Without such suggestions a number of important theorems would perhaps never have come to light. Would Daniel Bernoulli, for example[3] ever have thought that all periodic functions could be developed as a series of sine arcs that were multiples of each other, if his musician's ear had not, in the heart of each complex sound, discerned the simple sounds, the harmonics which compose it. Perhaps, in a world of the deaf, the Dalemberts, the Eulers, the Lagranges would never have imagined the trigonometric series, thus depriving analysis [*l'Analyse*] of one of its most comprehensive theories, and celestial mechanics and physics of one of their most potent aids.

So, Henri Poincaré rightly wrote:

> One must have completely forgotten the history of science in order not to recall that the desire to understand nature has had the most constant and the most felicitous influence on the development of mathematics.
>
> In the first place, the physicist poses for us the problems to which he expects the solutions from us. But in proposing them to us, he in large measure pays in advance for the service which we would render him if we succeed in solving them.
>
> If I might be allowed to continue my comparison with the fine arts, the pure mathematician who pays no attention to the existence of the exterior world would be like a painter who knew how to combine colors and forms harmoniously but to whom models were wanting. His creative power would soon be quite dessicated.[4]

When Poincaré used such language, it was on the strength of his own experience. Celestial mechanics and mathematical physics had set for him most of the problems in which his analytical genius had found such marvellous occasions for the exercise of its force and the proof of its fecundity.

From Descartes to Cauchy almost all the greatest mathematicians [*géométrés*] have been simultaneously great theoretical physicists. Consequently, they have taken care not to neglect this truth: Among the various sciences there is a natural order. By virtue of this order mathematical research departs from reality, in order to conclude in reality.

Concerned above all not to grapple with any problem incapable of resolution with ultimate precision, the German school of algebraists has not had the least concern for the natural sequence of human knowledge. Under the pretext of rendering mathematics more pure and more rigorous, this school has applied itself to wiping out from the [mathematical] sciences all that might recall their origins in mechanics or physics. Charles Hermite, for example, dealt with the theory of doubly periodic functions by procedures familiar to those who devote themselves to the study of electricity and magnetism. Weierstrass wanted to clothe that theory in a form in which the algebraic sequence was perfect, but from which the least resemblance to the methods of physics was banished.

There is no doubt that this misunderstanding of the place assigned to the mathematical sciences in the ensemble of human knowledge has been as damaging to mathematics as to physics. The former has lost in fecundity what the latter has lost in force and clarity, thereby gravely compromising the adequacy and the solidity of all science.

Up to this point we have been content with asserting that the mathematical mind was incapable, by its own resources, of establishing a natural order, be it in the domain of a single science or between various sciences. We must now show the reason for this incapacity. An example from botany will illustrate the principle that we propose to establish.

To put in order the multitude of plants, Linnaeus had proposed a systematic classification which was most readily put in place. You counted the number of stamens in a flower. According to whether you found one, two, three, four, etc., the flower would then take its place in a strictly delimited class. That class was, depending on the case, *monandrous, diandrous, triandrous, tetrandrous*, etc. Nothing could be more neat than this manner of classifying with a completely arithmetical simplicity. Nothing could do more violence to the natural affinities of plants. It happens, indeed, that plants, otherwise strongly analogous, do not bear the same number of stamens, and that the same number of stamens occurs in quite dissimilar flowers. Linnaeus foisted onto the plant kingdom a completely mathematical order, but not one that was in any way natural.

This fault struck Bernard de Jussieu. In 1758 Louis XIV had charged him with planting a botanical garden at the Trianon. Jussieu did not want the plants there to be grouped according to Linnaeus's artificial order. He wanted them to be arranged according to their natural analogies. The rules he applied were taken up again and perfected by his nephew, Antoine Laurent de Jussieu, who published in 1789 his *Genera plantarum secundum ordines naturales disposita.*[5] In this book, for the first time, botanists found a natural classification of plants.

The *Genera plantarum*, Cuvier said, "created in the observational sciences an epoch perhaps as important as the chemistry of Lavoisier did in the experimental sciences."

Laurent de Jussieu rejected "those arbitrarily constructed systems which offer us an artificial science rather than a natural one, which present us with a science condemned in advance to give us, concerning plants, no deep knowledge but only summary definitions and a certain manner of naming." The creation of similar systems cannot be other than provisional work, with which the mind contents itself "until such time as a repeated meditation shall have arranged plants, after a more felicitous fashion, according to a truly natural series."[6]

How then is this order of arranging plants according to their true affinities to be discovered?

> The characteristics of plants are not all of equal importance (*praestantia inaequales*). They are to be arranged in order, according to the dignity of the organ which they affect and according to the importance (*momentum*) of the various roles of this organ. Those characteristics which are inconstant or variable; those which are more stable; finally, those which are quite invariable or essential need not be used indiscriminately in the comparison of plants. They ought to be used conformably to this order.[7]
> However, this subdivision of characteristics into three classes does not suffice. Each of the classes admits of a

multitude of degrees which it is by no means easy to
define. To do so will be the pre-eminent task of the
botanist who assiduously studies nature to weigh attentively the importance of all the characteristics in order to
give to each of them the unchanging place to which it
belongs.—*Optimus labor botanici naturam sectantis is erit,
ut caracterum omnium momenta perpendat, suum singulis locum
daturus immutabilem.*[8]

In order to obtain a natural classification it does not suffice,
after Linnaeus's fashion, to arbitrarily choose a characteristic
which can be expressed in the language of arithmetic and to
operate simply on the basis of enumeration. It is necessary to
take all the characteristics and *weigh* them, so as to know which
exercise the greatest *moment*, and which produce less force upon
"the balance of affinities" (*affinitatum trutina*).[9] It is impossible
to state more clearly that the establishment of a natural classification transcends [*passe*] the powers of the mathematical mind
and that the intuitive mind alone is capable of attempting it.

The degree of importance, sometimes less and sometimes
more, is not, indeed, one of the notions which the mathematical
mind is capable of conceiving.

Several years ago, a plan of study, destined for I know not
what class, recommended that the professor of geometry demonstrate only the most important theorems. This occasioned much
hilarity among mathematicians. In a chain, no link is more or
less important than any other. Large or small, when a link snaps
the entire chain is broken. The author of this plan of studies
had imprudently extended to the domain of mathematics what
is true in the domain of intuition.

Is a given characteristic of an object an essential mark or
an accessory particularity? Is a given resemblance between two
beings a real and profound analogy or in fact an apparent and
superficial likeness? Ought a given doctrine be held as domi-
nant or subordinate? These are the things that can be intuited
but cannot be inferred.

The intuitive mind alone, then, can give a natural ordering to a science, because it alone can determine the degree of importance of diverse truths. This determination is necessary if it hopes to place in full view essential propositions, those by which the understanding will come to discern the function of less important propositions, and to discover the analogies which bind these latter to each another. These secondary propositions will be illuminated by the reflection of the splendor by which the principal propositions shine. Finally, the penumbra will envelop the details, insignificant as they may be.

The mathematical mind of Germany cannot conceive what we mean by insignificant detail.

A young German doctor had come to Pasteur's laboratory in order, he said, to become familiar with French microbiology. He was a student of Koch's. At Koch's 'institute' microbes were cultivated on slices of potato. Such was not the custom on the Rue d'Ulm. In order, no doubt, to know better the procedures of the latter laboratory, our German maintained he would do only that which was done in the former laboratory. Someone said to him: "That's no problem. Cultivate your bacilli as you prefer. Here are some potatoes." "But where is the knife to pare them?" "Take the first knife you come across, and if you don't find any, buy a pocket knife in the market for thirteen *sous*." "In Berlin, we have a special knife for paring potatoes." And our doctor would not begin his researches until he had received from Koch's laboratory that sacred instrument for paring potatoes. In such a fashion this knife came to rank as part of the scientific method.

Incapable of distinguishing what is of capital importance from what is insignificant detail, the German will not know how he ought to compose a work from the moment that the deductive method does not rigorously prescribe the order to follow. He will pay no attention to the art of setting essential ideas off by brilliant relief, and softening the light little by little in proportion as it falls upon thoughts of less value. Deprived

of the intuitive mind which would permit him to weigh
analogies and differences, he is unable to classify the subjects
he deals with in a natural manner. In the absence of natural
classification, he will search for some mathematical order for his
work, and the more rigid this order is, the more frequent and
starkly contrasting the dichotomies in it are, the more meticulous
subdivisions rarified to infinity there are in it, the more the
author will declare his satisfaction with it. But as this systematic
order is not drawn from the very nature of things, but is rather
dictated most frequently by the examination of some secondary
particularity, it will not be a guide which clears things up for
the mind of the reader. It is much more likely that its forced
accommodations, its pronounced separation of items with most
striking affinities, will provide a source of confusion. The inflex-
ibility of its mathematical style in matters which relate to the
intuitive mind gives it the air of wisdom. But it is only pedantry.

Furthermore, as detailed as the procedure may be, this
order will not penetrate to the ultimate detail of the science
which it presumes to present. Reduced as the headings may be
on which it gives up imposing its framework and subdivisions,
still everything must needs be brought to those ultimate elements.
Then, deprived of the intuitive mind and abandoned to the
mathematical mind, the author drowns himself in the most
inextricably grandiloquent nonsense imaginable.

In order to find striking types of this pathos to which the
German is condemned, to help him out of which he has no other
assistance than that of an artificial mathematical order, it almost
suffices to open at hazard a German treatise. I wish to cite only
one example. Camille Saint-Saens came across this example recently
in the works of Richard Wagner. It is a definition of melody.

> Melody is the redemption of indefinitely conditioned poetic
> thought, which is brought about by a deep consciousness
> of the highest freedom of emotion. It is the unwilled willed
> and accomplished, the unconscious conscious and proclaimed,
> necessity justified by an indeterminate content, condensed

from its most distant ramifications in view of a well-defined exteriorization of a content indefinitely extended.[10]

Wagner said one day to Frédéric Villot: "When I reread my former theoretical works, I cannot understand them."

When he no longer understands himself, the German is convinced that he has finally attained the heights of metaphysics. He has not grasped Voltaire's irony.

We have ended this analysis of German science. We have discovered its profound vices in the excessive development of the mathematical mind, in the miscarriage of the intuitive mind and even of simple common sense. They have prevented the German from producing the excellent fruits which his immense labor deserves.

Very often, no doubt, in following this analysis, you have thought that this contagion has overflowed to this side of our frontiers. Alas! It is only too true! For a long time now French science, forgetful of its glorious traditions, has set itself servilely to copy German science. Of this infiltration of the German spirit, of this slow poisoning of the genius of our nation, we do not wish at this time either to retrace the history, or search out the causes, or stigmatize the guilty authors. We wish to forbid ourselves all recrimination on the matter of the past. We have no desire to look back; we look forward.

Dear students, dear young people of France, you are preparing yourselves to restore to your country, at the price of your noble blood, the lands which others have stolen from it. When you will have accomplished this glorious work, there will remain yet another obligation for you to fulfill. That will be, by your work, to restore [*rendre*] to the fatherland the fullness and the purity of its soul. Together, then, let us search out how you will accomplish this task. Let us examine how you will protect your reason against the Germanic poison.

Will you ignore German science in the future? This is the advice which one hears often from voices possessing no authority. It is impossible to follow this advice. It would moreover be quite unfortunate if it were followed!

German experimental science and German erudition have amassed mountains of materials. It would be insane not to use them in building up the temple of truth. Of course, these materials, these observations, these texts, ought not to be received uncritically. It is important to make certain, by a rigorous examination, that excessive preoccupation with a preconceived idea has not introduced dishonesty, that the axiom, *Deutschland über alles,* has not mutilated and falsified them. But prudence is not abstention. In order to remake French science, you will draw heavily on the treasury of documents accumulated by German science. Like the Hebrews when they left the land of their servitude, you will carry away the gold vases of the Egyptians.

Neither will you shield your mind from every influence that comes from Germany, for, among the stimuli such influences can bring you, there are some that are excellent.

The absence of good sense and intuitive mind among the Germans is quite common. But there are numerous and quite happy exceptions to this general rule. There are German *savants* whose genius, perfectly balanced, knows how to allot to each faculty its just place and to use in turn the intuitions of common sense and the deductions of the geometrical mind. For example, which of the reproaches that we have addressed to German science could be applied to a Clausius or a Helmholtz? There is a school among the masters beyond the Rhine in which you can place your complete confidence. Your intelligence will draw nothing but profit from it.

There is still more to be said. The study of many a German work in which the excess of the mathematical mind completely effaces any trace of the intuitive mind can however possibly be quite profitable to you. From this mathematical mind, so heavy and so slow, you can borrow two eminently precious qualities which are too often lacking in us.

Our vivacity of mind [*esprit*] readily gives way to the counsels of the seductive imagination. We love to run, to fly toward any bright and distant object, without taking care to observe the precipices flanking the way. In the field of positive science as in the realm of history, reality ofteh appears less beautiful to us than fiction. The German mathematical mind will teach us patient rigor. It will teach us the art of proposing nothing for which we do not have the proof. Whenever some historical judgment was proposed before Fustel de Coulanges, the master would ask: "Have you a text supporting it?" Students were sometimes amused by the persistent regularity with which this anticipated question recurred. By this means the great French historian brought them back to that which is legitimate in the demands of the mathematical mind of the Germans.

The mathematical mind [*esprit*] is not simply the prudent mind. It is also the orderly [*de suite*] and tenacious mind. German science will teach you not to flit from one idea to another, like a butterfly from one flower to another, but, as patient and industrious bees, not to abandon a thought until you have· drawn from its nectaries all the honey that swells them.

You will not flee, then, from German influence. You will willingly receive all the salutary impulses that it can give you. But, from then on, you will need a sure means of not falling into the excesses it would threaten to drag you into. Who will procure this means for you?

Will it be supplied you by an opposite and rival influence, that of English thought?

Nothing is more opposite to German thought than English thought. In the latter, there is no desire for rigorous reasoning which could tie judgments to each other, no search for a systematic artificial order, in a word, no mathematical mind [*esprit*], but a prodigious power of clearly and distinctly seeing a multitude of concrete objects, all the while allowing each of them its own place in the complex and changing reality. Far from being excessively deductive, English science is all intuition. Nothing, then, it appears, is a better counter-balance to the exaggerated influence of German thought than the influence of English thought.

Be on your guard, though. Admire the English genius; do not imitate it. In order to proceed in the English fashion in the search for truth, you need the English mind [*esprit*]. You must possess that extraordinary faculty of imagining simultaneously a multitude of concrete things without feeling the need to arrange them, or classify them. Now, it is quite rare for a Frenchman to be endowed with this faculty. He has, on the contrary, an aptitude for conceiving abstract ideas, for analyzing them, for ordering them, which is lacking in the Englishman. In the conquest of truth then, the French and the English work each on his own side and after fashions appropriate to each. Both of them will obtain marvellous results, but let the one not be tempted to imitate the moves of the other, for that would only spoil things. Let the fish swim since it has fins, let the bird fly since it has wings. But do not advise the bird to swim or the fish to fly. We owe admirable discoveries to English physics. But the insane desire to copy that science has transformed the very harmonious and very logical theoretical physics composed by the French into a shameful and confused heap of illogicality and nonsense. Let this affirmation alone suffice for these lectures. Don't ask for the proof, since all recourse to the past has been excluded from them.

You will find yourself then subjected all at the same time to German and English influences, ready to receive from each of them whatever salutary impressions they may exercise, but resolved not to let yourself be seduced by either into the abyss wherein your French genius would founder.

It is possible to navigate in the strait of Messina, but it is necessary to have a clearsighted, strong-armed helmsman at the tiller, one who guards against every lurch toward Scylla or Charybdis. One can take poison and its antidote at the same time, but the doses must be quite strictly balanced. Where are you to find then the principle of integrity and balance of judgment which will guarantee your intelligence from the German peril as well as from the British? You will find it in the study of those who have regulated their reason in a perfectly straight-forward fashion, of those who have not allowed their reason to incline toward any excess. You will find it in the study of the classics of science.

Without respite, give over the care of forming your thought to those who were our precursors and our masters. Mathematicians, engineers, and astronomers, read Newton and Huyghens, Dalembert, Euler, Clairaut, Lagrange, and Laplace. Physicists, read Pascal, Newton, Poisson, Ampère, Sadi Carnot, and Foucault. Chemists, study Lavoisier, Guy-Lussac, Berzelius, J.-B. Dumas, Wurtz, Sainte-Claire Deville. Physiologists, meditate on Claude Bernard and Pasteur. Historians, take Fustel de Coulanges as your model. Nourish your mind with works in which the author has been able to make the proper distinction between the intuitive and the mathematical minds, where penetrating intuition has sensed the principles and a rigorous deduction concluded to the consequences.

But perhaps you will say that science has changed a lot since the time that these great men wrote. Science, yes, but not the manner of doing science, or at least doing it well. Do not believe those who repeat: 'We reason completely differently from, and better than, our ancestors.' In every age one comes across those presumptuous persons who affirm that, before them, the human intellect was in its infancy, and that with them alone it has left its apron-strings behind. This is a handy doctrine for those lazy ones whom it dispenses from studying the works of the past, or for the impudently vain whom it authorizes to give out old ideas as novelties. But it is a doctrine which collapses at the least glance over the history of the sciences. From Plato to our times the faculties which the human reason has at its disposition in order to search for truth have remained the same. And if our mind bit by bit brings to perfection the art of studying this or that subject it does so with extreme slowness and imperceptible progress.

It could be said to you, for example, that true method in the biological sciences only dates from yesterday. Take, however, the memoir on physiology in which, in 1651, Jean Pecquet of Dieppe, by his vivesections and painstakingly conducted experiments, set forth the laws of lymphatic circulation, and verified the laws of the circulation of the blood which anatomy had revealed a short while before to William Harvey.[12] Next to this short treatise put one of the well-wrought works of Claude Bernard. The two writings will appear to you to be contemporary with each other. During the two centuries that separate Bernard from Pecquet, physiological knowledge has developed remarkably; but the art of reasoning well in physiology has not changed.

Every time, then, that you wish to assist a science to progress directly and soundly, go to school with those who made the first steps in it.

I said to you, 'Read the classics of science'. I did not say to you, 'Read the French classics'. Far from me, indeed, is the thought of restricting to our country the glory of having produced works which ought to serve you as examples.

Assuredly, the men who have conducted their reason in a perfectly proper manner, who have maintained the most exact balance between their diverse faculties, have been more numerous among us than in any other land on earth. Foreigners agree with this willingly. They gladly cite rectitude and balance as the hallmarks of the French mind. But, for the good of the human intelligence, God has willed that no one nation should have the exclusive prerogative of these qualities. He has willed that each and every people should be able to discover with legitimate pride some geniuses among them in whom intuition and deduction should be developed with equal amplitude, and harmoniously proportioned.

Besides, there was a time when all scholars, formed by a similar discipline, had in view the same models, furnished them by the genius of antiquity. Then these men produced masterworks which were neither French nor Italian nor English nor German, but simply human. So, these qualities of mind [*esprit*] which I wish for you, and which you will come across in their highest degree in Frenchmen such as Descartes or Pascal, you will find again in Italians such as Galileo or Torricelli, in Englishmen such as Newton, Dutchmen such as Huygens, Germans such as Leibniz, and Russians [sic] such as Euler. And if I had to cite a perfect example of clarity, good sense, order and measure, I would find it in the mathematician and physicist whose motto was *Pauca, sed perfecta*—the German, Carl Friedrich Gauss.

Read then the classical authors, all the classical authors. They will restore to you those two qualities which were long the mark of the French mind and which, alas, we have completely abandoned: clarity and good sense.

Clarity! How often in my youth, did I hear people making fun of it! Under the influence of masters blinded by German prestige, we had come to that aberration of confounding obscurity with profundity. We poked fun at the verses of Boileau:

That which is well-conceived is clearly articulable.

People claimed the right to speak obscurely about obscure things. No! A thousand times, no! There is no right to speak of an obscure thing except to clarify it. If the only effect of your verbiage must be to confuse things further, be still!

French students, beware of those who would accustom you to think confusedly [*dans la nuit*]. From always hunting in the dark, the owl ends by not being able to see in broad daylight. From meditating incessantly in the German mists, some have become incapable of comprehending that which is clear. "Too much truth astonishes us," said Pascal. "I know some who cannot comprehend that four from four gives zero."[12] Flee from these intellectual owls who would wish to make you dazzled like them. Accustom your eyes to look the splendor of the truth in the face. In all cases, I conjure you, be intransigent defenders of clarity. When you come across one of these philosophers or physicists who is satisfied to live in fog and confusion, do not allow him to claim profundity. Lift up the mask which covers his ignorance and sluggishness of mind. Say to him simply: "My friend, if you do not succeed in making us understand what you are talking about, it is because you yourself do not understand it at all!''

Become the defenders of clarity. Make yourselves and those around you the defenders of good sense.

Good sense you will hear quite frequently denigrated, and, perhaps, you will be tempted to lend a too willing ear to such denigration.

You will hear it said that good sense is the enemy of originality. Do not, for pity's sake, apply the good word 'originality' to those absurdities, oddity and extravagance.

In Paris I lived once for a whole year in a small garret room. From my window, beyond many chimneys, the tower of Saint-Jacques du Haut-Pas could be seen, and the top of a huge tree in whose shade Malebranche had meditated.

On the floor immediately below lived a family in bourgeois comfort. A little man, about 65 years of age, graying, neat; an attack of hemiplegia some time before had made his step shuffling, his hand clumsy, his mood peevish and at times irritable. His companion was a model of solicitous activity and perpetual devotion. She had as her constant preoccupation to keep even the least concern from the husband God had given her. It was impossible to imagine a life more simple or integrated, or people who were more like everyone else, or at least as everyone else ought to be. Nevertheless, when they went out, always together, passers-by stopped to look at the halting gait of the old man. Then, when he had passed by, you could hear them murmur with veneration: 'Pasteur!' It was, in fact, the scientist [*savant*] who had just crowned his glorious career with the discovery of the vaccine against rabies.

No man was more commonplace than Pasteur. Would you deny him originality?

It will also be said further: Where good sense reigns as master, there is no more poetry! And what accusation could appear more serious to your souls when you are twenty years old?

Let my memories flow before you again.

Not very long ago, in Provence, I went with pleasure to return a visit to a venerable neighbor. A short bit of road along the *roubines* covered with roses, in the shade of rows of huge cypress trees in the intervals between which one could see in the distance the Sphinx-like rocks of Saint-Rémy and the *Alpilles* with their elegant indentations, brought me to the gate of this fine old man. His conversation was for me a real feast. In harmonious language, sprinkled with neat and sober images, he spoke to me of country things and country people. He always judged them with the greatest kindness, but also with the most shrewd penetration. He had, in his time, conceived major projects—not just vague dreams but plans examined with mature consideration, the execution of which he had followed with as much ability as perseverance. I hailed in him the perfection of French good sense, an accomplished example of the intuitive mind.

From his youth, however, he had applied himself to poetry. Now, his first verses had drawn this cry from Lamartine: "I am going to tell you today a bit of good news: A great epic poet is born to us!"[14] For my old country neighbor, so judicious and so shrewd, was the singer of Mireille and of Esterelle, of Nerte and of the Anglore. He was the Provençal poet whose voice, from Ventour to Canigou, made all the echoes of the *Langue d'oc* dialects vibrate. He was Frédéric Mistral.

From the day I first knew Frédéric Mistral, I understood the words of Horace:

Scribendi recte sapere est et principium et fons.

The principle and the source of great, of immortal, poetry is good sense.

To receive from foreign influences, from the English as well as the German, all the salutary impulsions, but to guard yourself at the same time from all pernicious seduction, you should maintain your reason in profound respect, in the continual practice of this good sense and this clarity which were, with us, matters of tradition. Your good sense will be applied to discern with great exactitude the false from the true in everything. And when you shall have made this discrimination, in all simplicity, in all loyalty, in full clarity, you shall say to truth: Yes, you are; and to falsity: no, you are not. *Sit lingua vestra: Est, est; non; non.* The divine master has said it. It is thus that you must think, that you must speak, if you want your thought and your speech to be Christian. But when, inside as well as outside, your word shall conform to this rule, it will be frank [*il sera franc*]: that is to say that you will think, that you will speak, in French [*en Français*].

Some Reflections on German Science

'Quelques réflexions sur la science allemande', *Revue des deux mondes*, 1st February 1915.

I

Formerly, we tried to describe the stamp which impresses a quite particular and salient character upon English theoretical physics. Today we want to try, in a similar manner, to lay bare the hallmark of the doctrines of mathematics or physics manufactured in Germany.

Such an attempt must guard against claiming to come to any rigorous conclusions. Taken in its essence, considered under its perfect form, science ought to be absolutely impersonal. Since no discovery in it would bear the signature of its author, neither would anything allow one to say in what land the discovery saw the light of day.

But this perfect form of science could not be obtained [ne raurait (sic) être obtenu] without a quite exact sorting out of the various methods which come together to reveal the truth. Each of the many faculties which human reason puts in play when it wishes to know more, and better, would need to play its role, without any being omitted, without any being overtaxed.

We do not come across this perfect equilibrium between the many organs of reason in any one man. In each of us, one faculty is more powerful, and another more weak. In the conquest of truth, the weaker faculty will not contribute as much as it should, and the more powerful faculty will take on more than its share. The science produced by this poorly apportioned work will not show the harmonious proportions of its ideal exemplar. With the faulty development of certain parts goes the excessive growth of others. It is in these deformities alone that we can recognize its author's turn of mind.

It is these too that will often allow us to name the nation that produced a particular theory.

The body of each man deviates from the ideal type of the human body by the exaggerated proportions of this organ, by the diminished proportions of that one. These sorts of modified monstrosities that distinguish us one from another are also those which characterize physically the various nations: one sort of exaggerated or arrested development is especially frequent among a particular people.

That which is said of the body can be repeated of the mind [*esprit*]. To say that a people has its own particular mind [*esprit*] is to say that, quite frequently, in the reason of those who shape this people, a certain faculty is developed more than is suitable and that a certain other faculty does not have its whole amplitude and all its force.

From that two conclusions follow immediately:

In the first place, judgments about the intellectual form of a people will frequently allow of verification, but they will never be universally applicable. All the English are not of the English type. With greater reason, the theories conceived by the English will not show all the characteristics of English science. Among them one will come across those that might well be taken for French or German works. In return, there will be found, among the French, intellects which think in the English mode.

In the second place, if the national character of an author is perceived in the theories he has created or developed, it is because this character has shaped that by which these theories diverged from their perfect types. It is by its shortcomings, and by its shortcomings alone, that science, distanced from its ideal, becomes the science of this or that nation. One can expect, then, that those marks of the genius proper to each nation will be particularly outstanding in works of the second order, the products of mediocre thinkers. Quite often, the great masters possess a reason in which all the faculties are so harmoniously proportioned that their quite perfect theories are exempt from every individual and even national character. There is no trace of the English mind [*esprit*] in Newton, nothing of the German in the work of Gauss or Helmholtz. In such works one no longer divines the genius of this or that nation, but only the genius of humanity.

II

"Principles are intuited; propositions are inferred," said Pascal,[1] whom we must always cite when we claim to speak about scientific method. In every science that has taken the form that we might call rational, or better yet, the form that might be called mathematical, we must, in fact, distinguish two strategies: that which seizes on [*conquiert*] principles, and that which arrives at [*parvient*] conclusions.

The method that, from principles, ends with conclusions, is the deductive method, followed with the most rigorous precision.

The method that leads to the formulation of principles is much more complex and difficult to define.

Is it a question of purely mathematical science? Common experience is the material from which induction draws the axioms. Deduction will draw out all the truths such universal propositions contain. Now, the choice of axioms is an operation of extreme delicacy. The axioms must suffice to justify all the propositions of the science that we wish to extract from them. The chain of reasoning must not suddenly have its continuity broken and its rigor compromised because a principle necessary for its progress has remained hidden in the data of experience and has not yet been formulated explicitly. It is equally necessary that there not be too many principles, that a simple corollary of an axiom be not given as itself an axiom. If we follow the history of the axioms of geometry from the *Elements* of Euclid to the works of Hilbert we will see how much the choice of the principles of a mathematical science is a minute and complicated task.

More complex yet is the choice of hypotheses upon which will rest the entire edifice of a doctrine pertaining to experimental science, of a theory of mechanics or physics.

Here the matter which ought to furnish the principles is no longer common experience, spontaneously available to every man from the time he leaves infancy. It is scientific experiment [*expérience*]. To the mathematical sciences common experience furnishes autonomous, rigorous, definitive data. The data of scientific experiment are only approximate. The continual improvement [*perfectionnement*] of instruments increasingly modifies them, while the fortunate chance of discovery each day comes to enlarge the treasury with some new fact. Finally, far from being autonomous, or immediately intelligible in themselves, the propositions which formulate the results of an experiment in physics or chemistry only have meaning if the accepted theories supply translations for them.

The physicist must extract his principles from this inextricable network where the data of sensation lie tangled up. In this he is assisted by more or less complicated instruments, by interpretations furnished by changeable theories subject to doubt, sometimes by the very theory he proposes to change. In the inspection of this confused mixture, he ought to divine the general propositions by which deduction will proceed to conclusions conformable to the facts.

He would find in the deductive method assistance that is just too rigid and not penetrating enough for the accomplishment of this work. He needs a more supple and subtle method than that. More so than the mathematician, the physicist, in order to choose his axioms, will need a faculty distinct from the mathematical mind [*esprit géométrique*]. He will have to appeal to the intuitive mind [*esprit de finesse*].

III

The intuitive mind and the mathematical mind do not proceed at the same pace.

The progress of the mathematical mind obeys inflexible rules which are imposed on it from another source. Each of the propositions which it unfolds one after another has its place marked out in advance by a necessary law. To evade this law, be it by ever so little, and to pass from one judgment to another by jumping over some intermediary required by the deductive method, is, for this mind, to lose its force, which consists entirely in its rigor. The word *enchaînment* comes to our lips as soon as we want to define the order in which the syllogisms succeed one another. In fact, the chain which ties together such reasoning allows for no liberty.

If the mathematical mind owes to the rigor of its approach all the force of its deductions, the penetration of the intuitive mind belongs entirely to the spontaneous suppleness with which it moves. No unchangeable principle determines the path which its free endeavors will follow. At one moment we see it, with an audacious leap, clear the abyss which separates two propositions. At another it slips into and insinuates itself among the many objections impeding the approach to a truth. It is not that it proceeds without order, but that the order it follows it prescribes for itself. It incessantly modifies the order in light of the circumstances and occasions in such a fashion that no precise definition could pin down its sinuosities and unforeseeable leaps.

The advance of the mathematical mind calls forth the idea of an army marching past on parade. The various regiments are aligned with impeccable regularity. Each man holds the exact position allotted to him by a strict order. He feels held there by an iron discipline.

The progress of the intuitive mind calls to mind rather the idea of sharpshooters sent to assault a difficult position. At one moment it leaps suddenly. At another, it creeps stealthily as it climbs through the obstacles with which the slope bristles. There, also, each soldier follows orders. But no part of the orders is explicitly formulated except the end of taking the position. The free interpretation of how to implement this goal made by each of the assailants tends to co-ordinate the various movements which to each seem most favorable to the specified goal.

Doesn't this comparison between the movement of the intuitive mind and that of the mathematical mind allow us already to predict the specific character of German science, that which will distinguish it in particular from French science? There is no doubt that, with the greater number of Frenchmen who cultivate it, science will be marked by an excessive application of the intuitive mind. Not content with the role which has devolved upon it, impatient of the slow ponderousness of the mathematical mind, the intuitive mind at times encroaches upon the prerogatives of the latter. We must also note that German science is often deficient in the intuitive mind and gives to the mathematical mind that which is not its legitimate possession.

Let us glance over some of the works which have established the reputation of German science and see if the predominance of the mathematical mind over the intuitive mind is not easily recognized here.

IV

The mathematical mind could still better be called the algebraic mind. There is no part of science, in fact, in which the deductive method has a greater role than in this vast generalization of arithmetic to which is given the name of algebra or analysis. The axioms which it rests on consist in a very small number of quite simple propositions concerning whole numbers and their addition. The intuitive mind did not have to make any great exertion in order to disengage them from the most common experience. From these axioms, by following the most rigorous syllogisms that it is possible to conceive, the innumerable truths of which the science of algebra is made can be drawn.

The faculty of following without lapse the scrupulously careful rules of logic in the course of long and complicated processes of reasoning is not, however, the only faculty which comes into play in the construction of algebra. Another faculty takes an essential part in this work. It is that by which the mathematician, in the presence of a very complex algebraic expression, easily perceives the various transformations, allowable by the rules of calculation, which he can make it undergo and by which he can reach the formulas he wanted to discover. This faculty, quite analogous to that of a player of checkers who prepares a clever move, is not a power of reasoning but rather an aptitude for combining.

Among German mathematicians there are doubtless those who have possessed this aptitude for combining the operations of algebraic calculation in a high degree. But it is not by this that the analysts from beyond the Rhine have excelled. Grand masters of this art are more easily found in France, and above all in England, such as a Hermite, a Cayley, a Sylvester. It is by its power of deducing with the most extreme rigor, of following without the least lapse the most extended and most complicated chains of reasoning, that German algebra has marked its superiority. It is by this power that a Weirstrass, a Kronecker, a Georg Cantor have displayed the force of their mathematical minds.

By this absolute submission of their mathematical minds to the rules of deductive logic, German mathematicians have quite usefully contributed to the perfection of analysis. The algebraists who had shone among other peoples before had too readily and more than properly trusted the intuitions of the intuitive mind. So, they had often come to formulate as demonstrated truths what were in fact only guesswork. Sometimes even propositions had been hastily given as true when they in fact were not. German science has greatly contributed to disencumbering the field of algebra of all paralogisms.

To cite only one example out of a thousand: By means of a quite ready and hasty intuition, the intuitive mind had believed it knew that every continuous function admits of a derivative.

Hurrying on the mathematical mind more than properly, it had come to accept from the latter some apparent demonstrations of that proposition. In forming continuous functions which never have derivatives, Weierstrass demonstrated how dangerous the momentary abandonment of rigor could be in the course of an algebraic deduction.

So the extreme rigor of the mathematical mind has great advantages for the progress of algebra. It also presents very serious inconveniences. Solicitous to excess of avoiding or resolving objections which are only trifles, it encumbers science with otiose and tedious discussions. It suffocates the spirit of invention. In fact, before forging the chain which ought to connect a new truth with principles by means of tried and tested links, it is indeed necessary in the first place to have caught sight of that truth. The intuition that precedes demonstration in every mathematical discovery is a prerogative of the intuitive mind. The mathematical mind does not know it, and, in the name of rigor, it readily denies to it the right to function. Uneasy about the dangers which exclusive usage of the mathematical spirit causes for the faculty of invention, there are even in Germany certain mathematicians such as Felix Klein who have come to assert the place of the intuitions proper to the intuitive mind in the domain of algebraic method.

V

Algebra subjects reason to this iron discipline consisting in the laws of the syllogism and the rules of calculation [*calcul*]. No science is better adapted to the German mind, proud of its mathematical rigor, but deprived of intuition. So the German has tried to give to every science a form which, as much as possible, resembles that of algebra. For example, in German hands, geometry was reduced to nothing but a branch of analysis.

Already, by the invention of analytic geometry, Descartes had reduced the study of figures marked out in space to discussion in terms of algebraic equations. Each point in space can be expressed by three numbers, the *co-ordinates* of this point. For a given point to be found on a particular surface, it is necessary and sufficient that its three co-ordinates satisfy a certain equation. All information concerning the algebraic properties of the equation is, *pari passu*, information about the geometrical properties of the surface, and inversely. He, then, who is more skilled at combining the formulas than at considering assemblages of lines and surfaces, will be a great geometer by virtue of the sole fact that he is a skilled algebraist.

Nevertheless, even after the work of Descartes, the reduction of geometry to algebra was not absolute. In order to assign three co-ordinates to a point in space, it was still necessary to have recourse to some geometrical propositions, to the most elementary theories about straight lines [*droites*] and parallel planes. Simple as these propositions may be, they entailed acceptance of all the axioms for which Euclid, at the beginning of the *Elements*, claimed acceptance. Now for some people, in whom the mathematical mind suffers from the least diminishment of rigor, this adherence to the axioms of Euclid is scandalous.

The axioms that a science of reasoning demands that we grant to it ought not merely to agree among themselves without any shade of contradiction. They ought, further, to be as few in number as possible. Consequently, they ought to be independent one from another. If one among them, in fact, could be demonstrated by means of the others, it would be deleted from the number of the axioms and relegated to the class of theorems.

Now, are the axioms of Euclid truly independent one from another? This is a question which has long disquieted geometers. Among these axioms there is one, that upon which the theory of parallel lines rests, which many have thought to recognize as a simple corollary of the other proposals formulated by Greek geometry. So we have seen repeated attempts at the demonstration of this postulate of Euclid. But the slightly perspicacious critic has always discovered a vicious circle in every one of these attempts.

More ingeniously, the question was given another slant by Gauss, Bolyai, and Lobachevsky. These mathematicians applied themselves to laying out the series of propositions which could be established by accepting all the axioms formulated by Euclid, except the parallel postulate. If, they thought, one can follow out to infinity the series of consequences of those axioms, without assuming the truth of the contestable postulate and without ever, however, coming up against a contradiction, it is consequently the case that the adoption of these principles would not necessarily require the truth of that postulate which supports the theory of parallels. Henri Poincaré has shown the complete cogency of this thought conceived by Gauss, Bolyai, and Lobachevsky. He has demonstrated that if the non-Euclidean geometry constructed by these mathematicians could ever lead to two mutually contradictory propositions it would be because Euclidean geometry itself could furnish two incompatible theorems.

To find out whether all the axioms of Euclid are truly independent of each other is a question under the jurisdiction of the mathematical mind. And, with Gauss, Bolyai, Lobachevsky, and their successors that mind has fully resolved it. But to decide whether the postulate of Euclid is true is a question that the mathematical mind, left to itself, could not answer. It must, in this case, have recourse to the aid of the intuitive mind.

The truth of geometry does not consist merely in the absolute mutual independence of the axioms, or in the impeccable rigor with which the theorems follow from the axioms. It consists also and above all in the agreement between the propositions which form this logical chain and the knowledge given to our reason concerning space and the figures which can be traced in it, by that extended experiment called common sense. It belongs to the mathematical mind to verify the exactitude of the deduction by which all the propositions are drawn from each other [*se tirent les uns des autres*]. But it has no means of recognizing if these are or are not conformable to what we know, prior to all geometry, about plane or solid figures. This latter concern is the task of the intuitive mind.

Now one of the first truths which we are capable of formulating on the subject of space, prior to all geometry, is that space has three dimensions. When the intuitive mind analyzes this proposition in order to grasp exactly what is meant by the formulation, does it discover that it has this sense: To each point in space there correspond three numbers which are its co-ordinates? Not at all. What it finds is that, in attributing three dimensions to space, the man who is not a mathematician claims to say this: All bodies have length, breadth, and depth. And if it presses this affirmation, the intuitive mind recognizes that it is equivalent to this other affirmation: Every body can be exactly contained in a box [*boîte*] of precisely determined size, whose shape is called by the geometer a rectangular parallelipiped. The mathematical mind then comes along to demonstrate that propositions relative to rectangular parallelipipeds, judged true by the intuitive intellect, entail the celebrated postulate of Euclid.

Furthermore, in searching about in the treasury of truths relative to sizes and figures which the most common experience stores up, the intuitive mind comes across these propositions: By drawing it is possible to represent a plane figure, by sculpture a solid figure, and have the image exactly replicate the model, although it be a different size. This is a truth which was not in any way doubted in paleolithic times by the hunters of reindeer from the banks of the Vezere. Now, that figures can be similar without being equal presupposes, as the geometrical mind demonstrates, the correctness of the postulate of Euclid.

To recognize in this fashion the highly significant role devolving to the intuitive mind in the verification of the axioms of geometry is not to the taste of German science. The latter will make little of the accord between the propositions of geometry and knowledge drawn from common sense, since this accord cannot be established by the mathematical mind. It will have the truth of geometry consist exclusively in the rigor of deductive reasoning by which theorems derive from axioms. And, in order not to be exposed to compromising this rigor by deriving some information from sensible experience, it would reduce geometry to absolutely nothing more than a problem of algebra.

For German science, a point will be *by definition* the ensemble of three numbers. Let the values of the three numbers vary continuously in such an ensemble, then it will be said that the point generates a space. The distance between two points will be *by definition* an algebraic expression in which there appear the three numbers of the first ensemble and those of a second. This algebraic expression will certainly not be taken absolutely at random: it will be chosen in such a manner that some of its algebraic properties express themselves by phrases analogous to those which articulate certain geometrical properties attributed by common sense to the distance between two points. But we will seek to make such properties as few as possible, for fear that the intuitive mind may find in them a pretext for penetrating into the domain of the science one wishes to construct. Then, algebraic calculations will be developed and called geometry.

Perhaps the intuitive knowledge reason furnishes us concerning plane figures and bodies would still find means of insinuating itself into the interstices of the deductive net which this algebra weaves. Against this redoubtable intuition, a new precaution will be taken. Such intuition knows no point in space which does not have two or three dimensions. To articulate propositions which might speak of a space of more than three dimensions would be to speak words which, to such intuition, are senseless. These are precisely the propositions we will constantly attempt to formulate. That which will be called a point will not be, as we have thought, an ensemble of three numbers, but an ensemble of n numbers. The value of the whole number represented by n will not be specified. This value can be more than three; it can be as great as we wish. This ensemble of n numbers is, it will be said, a point in a space of n dimensions.

In this fashion did the powerfully mathematical genius of Bernhard Riemann go about writing a chapter of profound algebra to which he gave the title: *On the Hypotheses Which Serve as the Foundations of Geometry (Über die Hypothesen welche der Geometrie zu Gründe liegen).*

We have pointed out with what minute care the intuitive knowledge of lines and surfaces had been held in abeyance in the composition of that theory. Is it astonishing that the corollaries to which that algebra leads, and which it sets forth with words borrowed from geometry, run headlong against the propositions which the intuitive knowledge of space regards as the most certain? When it affirms, for example, the meeting at a finite distance of any two lines whatsoever in the same plane, doesn't it deny the very existence of parallels?

The theory of Riemann is a *rigorous algebra*, for all the theorems which it formulates are quite exactly deduced from postulates which it sets forth. Hence it satisfies the mathematical mind. It is not a *true geometry*, for, in posing its postulates, it does not take care that their corollaries should accord in every point with judgments drawn from experience which compose our intuitive knowledge of space. It thereby shocks common sense.

VI

Riemann's memoir on the foundation of geometry is one of the most justly celebrated works of German science. It appears to us to be a remarkable example of the procedure by which the mathematical mind of the Germans transforms every theory into a sort of algebra.

This mind assigns extremely unequal shares to the two methods by whose assistance every science of reasoning progresses. It develops with as much amplitude as detail the process of deduction by which the corollaries are drawn from principles. It suppresses or reduces to the slightest importance the ensemble of inductions and divinations by which the intuitive mind was able, on the basis of experience, to disengage principles.

The hypotheses upon which any theory whatsoever of mechanics or mathematical physics rests are fruits whose maturity has been prepared over a long period of time. Data of common observation, results of scientific experiment assisted by instruments, ancient theories now forgotten or rejected, metaphysical systems, and even religious beliefs have contributed to them. Their effects are so intersected, their influences so mixed in so complex a manner, that a great subtlety of mind, sustained by a profound knowledge of history, is required in order to discriminate the essential direction of the path that has led human reason to the clear perception of a principle of physics.

Now, let us examine some of the lessons of a most scientific algebra, in which Gustav Kirchoff has set forth the diverse theories of mathematical physics. Of that long and complicated elaboration that preceded the adoption of principles we find no trace. Each hypothesis is presented *ex abrupto*, under the quite abstract and general aspect which it has taken on after many evolutions and transformations, without any word that might make us suspect the indispensable preparation. A Frenchman who had been an auditor of Kirchoff's in Berlin repeated to me not long ago the formula by which the German professor was accustomed to present each new principle. "We can and we will posit [*poser*] . . . *wir können und wollen setzen* . . . " Provided that any contradiction does not forbid the pure logician the use of the supposition which we want to make, we prescribe it as a decree of our free will. This act of will, this choice of our good pleasure, is substituted, so to say, purely and simply, for all the work that, over the course of the ages, the intuitive mind had to complete. It doesn't leave anything more subsisting in science than that which submits to the rude discipline of the mathematical mind. A theory of physics, based on postulates freely formulated, is no more than a series of algebraic deductions.

Kirchoff is not alone in dealing in this fashion with mechanics and physics. Those who have followed his lectures imitate his

methods. Is it possible to imagine, for instance, a more absolute
algebraism than that which inspires Heinrich Hertz when he claims
to construct mechanics? The disposition, at a given instant, of
the various bodies composing the system studied is known when
one knows the values taken by a certain number of magnitudes,
n. For fear that experimental intuition might come to suggest to
us some property of this system of mechanics, we quite quickly
drop from sight and forget the bodies which form the system,
outwit intuition, and consider only a point whose co-ordinates,
in a space of n dimensions, will be precisely these n values. Let
us agree that this point—which is itself nothing but an algebraic
expression, only a word of geometric consonance taken to designate
an ensemble of n numbers—changes, from one instant to another,
in such a fashion as to minimize a certain magnitude, represented
by an algebraic formula. From this convention, so perfectly
algebraic in nature, so completely arbitrary in appearance, we
deduce, with perfect rigor, the consequences that calculation
can draw from it, and we say that we are setting forth mechanics.

Certainly, the postulate formulated by Hertz is not as arbitrary as it appears. It has been disposed in such a fashion that
its algebraic articulation summarized and condensed all that—
from Jean Buridan to Galileo and Descartes, and from the latter
to Lagrange and Gauss—intuition, experiment, and discussion
had disclosed to mechanicians concerning the law of inertia and
the connections by which bodies mutually inhibit each other in
their motions. But of all that previous elaboration Heinrich Hertz
has not preserved the least reminder in the absolutely precise
and rigorous exposition of mechanics which he gives us. He
completely and systematically eliminates it, so that the fundamental principle of science takes the imperious form of a decree
issued by an openly authoritarian algebraist: *Sic volo, sic jubeo,
sit pro ratione voluntas.* [I will it thus, I order it thus; let my will
stand in the place of reason.]

Such a manner of proceeding can, in certain cases anyway,
produce quite felicitous consequences.

By constant unravelling of the complex skein of operations which have slowly produced a hypothesis in physics, the intuitive mind at times deludes itself concerning the role that it plays. It comes to imagine that it has accomplished a work of the mathematical mind. The series of considerations, with transitions delicately arranged, through which it has little by little prepared the mind [*esprit*] to receive a proposition, it wrongly takes for a categorical demonstration of that proposition. Our French physics has too often and too long given itself over to this illusion. It is important to put reason on guard against this misapprehension, not to allow it to believe that a principle of physics is demonstrated when we have only rendered it captivating. It is good to remind reason that, from the point of view of deductive logic, the hypotheses of physics present themselves under the aspect of propositions which no reasoning imposes; that the scientist [*savant*] formulates them as he pleases, led solely by the hope of drawing from them corollaries conformable to the data of experience; that he proposes them for our acceptance because the condensation of a multitude of experimental laws and a small number of theoretical postulates appear to him, in the words of Ernst Mach, a fortunate economy of thought. For this task the pure algebraism of German theories is marvellously apt.

But what is to be said? Simply that an exposition of physics in which the intuitive mind had exaggerated its power is corrected by another exposition in which that mind has been driven out with too much brutality. In other words, that excess in one direction often finds its remedy in excess in a contrary direction. Each is nonetheless an excess. Belladonna and digitalis neutralize each other's effects. They are nevertheless two poisonous plants.

VII

In positing the hypotheses of a theory of mechanics or physics without any concern for the considerations by which the intuitive mind could prepare for our acceptance of them, one takes the risk of giving in to a serious failing. One opens oneself to the production of doctrines which shock the universally received teachings of common sense.

German science cares little for the exigencies of common sense. It is not displeased to oppose it directly. The geometric theory of Bernhard Riemann has already allowed us to recognize this. At the base of the systems it constructs with an apparatus so scrupulously designed, German thought appears at times to take a malicious pleasure in positing some affirmation which, for the intuitive mind, may be the occasion of scandal, even if this affirmation should contradict the most assured principles of logic. To posit, then, a proposition formally contradictory to a number of axioms—and by a series of quite conclusive syllogisms to draw a whole ensemble of corollaries therefrom—what a delicious exercise for a mathematical mind which despises the intuitive mind and good sense!

From an early period men were found in Germany to uphold this temerity.

Before the middle of the fifteenth century, Nicholas of Cusa, the first original thinker of whom German reason can boast, wrote his treatise: *De docta ignorantia.* In order to provide a basis for the philosophical edifice which he was going to erect, the 'German cardinal' posed this affirmation, the contradictory character of which leaps to one's eyes: In every order of things, the maximum and the minimum are identical. Then, on this foundation, the deductive method allowed him to construct an entire metaphysics.

The nineteenth century produced, in Germany, an attempt no less strange than that of Nicholas of Cusa. Hegel came to rest his entire philosophical system on the affirmation of the identity of contraries. And the great success which Hegelianism has known in universities beyond the Rhine marks to what extent the mathematical mind of the Germans, far from being shocked by this defiance of common sense, has taken pleasure in that *tour de force* of the purely deductive method.

A being whose nature consists in being conscious of being dominated by an iron discipline finds its happiness in obeying without discussion the order which it obeys. The more strange— even revolting—this order may be, the more is obedience joyful for such a person. Thus is explained the ready obedience with which the mathematical mind of a Nicholas of Cusa or a Hegel rolls out the consequences of an absurd principle. The metaphysicians, moreover, have not been the only Germans to supply examples of this intellectual submission which disconcerts us. We have seen mathematicians spin off entire geometries in which some one of the least discussable of the axioms Euclid had formulated was replaced by its contradictory. And the authors of these deductions appeared to take pleasure in direct proportion to the inconceivability and preposterousness of the conclusions to the judgment of good, old, common sense.

It is nevertheless a geometry conforming to good old common sense which these mathematicians use whenever, in everyday practice, they measure any body or draw any figure.

A similar inconsistency is not rare in instances where the mathematical mind claims to do without the assistance of the intuitive mind. Isolated from common sense, the mathematical mind can reason fluently and deduce endlessly. But it is incapable of directing action and of maintaining life. It is common sense which reigns as master in the domain of facts. Between this common sense and discursive science it is the intuitive mind that establishes a perpetual circulation of truths, that extracts from common sense the principles from which science will deduce its conclusions, that recapitulates among its conclusions all that can enhance and bring to perfection common sense.

German science does not know this continued exchange. Submissive to the rigorous discipline of the purely deductive method, theory follows its regular march without any concern for common sense. Common sense, on the other hand, continues to direct action, without theory coming by any means to hone away its primitive and gross form.

Doesn't the idealist philosopher crudely display this absence of all interpenetration between science and life? In his chair at the university, he denies all reality to the exterior world, because his mathematical mind has not come across this reality at the end of any conclusive syllogism. An hour later, at the tavern, he finds a fully assured satisfaction in those solid realities, his sauerkraut, his beer, and his pipe.

With the Germans, pure mathematicians deprived of the intuitive mind, life does not guide science and science does not enlighten life. Hence, in his magnificent work on *Germany and the War*, M. Émile Boutroux could write:

> Their science, an affair of specialists and scholars, has not been able to penetrate their soul and influence their character. . . . Apart, to be sure, from notable exceptions, consider this learned professor in the tavern, in his relations of ordinary life, in his amusements—he who excels at discovering and gathering together all the materials for a study and who has brought forth from it, through mechanical operations and without the least appeal to judgments and good common sense, solutions completely based on texts and processes of reasoning. What a disparity there often is between [his social behavior and] his science and the degree of his education! What vulgarity of taste, sentiment, language! What brutality of conduct with this man, whose authority is inviolable in his specialty! . . . with the Germans, the scholar and the man are only too often strangers to one another.[2]

It is with German science as with the German scientist. The absence of the intuitive mind leaves a gaping abyss there between the development of ideas and the observation of facts. Ideas are deduced one from another, proudly contradicting the common sense to which they owe nothing. Common sense manipulates realities and sets forth facts by its own proper means, without concern for a theory which ignores it or comes into collision with it. Such is the spectacle which, quite often, is presented to us today by physics from beyond the Rhine.

VIII

German theories of electric phenomena will offer us examples of this incoherent duality.

There is, in mathematical physics, a particularly difficult and complicated theory, namely, the theory of electricity and magnetism. The genius of a Poisson, of an Ampère, set the principles of this doctrine with characteristically French clarity. The work of these great men had, before the middle of the nineteenth century, served as a guide for the work which the most illustrious German physicists—Gauss, Wilhelm Weber, and Franz Neumann—had carried out in order to complete it. All these efforts, inspired by the intuitive mind and at the same time disciplined by the mathematical mind, had built up one of the most powerful and most harmonious theories of physics that had ever been admired. For several years now this doctrine has been thrown into complete confusion by the exclusively mathematical spirit of the Germans.

The origin of this confusion, however, does not lie in Germany. Its origin must be sought in Scotland.

The Scots physicist James Clerk Maxwell was as if haunted by two intuitions.

The first of these was that insulating bodies, those which Faraday called *dielectric,* ought to play a role in regard to electrical phenomena comparable to that which conducting bodies play. It is appropriate to constitute an electrodynamics for dielectric bodies analogous to that which Ampère, W. Weber, and F. Neumann constituted for conducting bodies.

The second was that electrical movements ought to propagate themselves, in the heart of the dielectric body, in the same fashion as light is propagated in the heart of a transparent body. And in the same material, the velocity of electricity and the velocity of light ought to be the same.

Maxwell sought, then, to extend to dielectric bodies the equations of the mathematical theory of electricity and to put these equations into a form such that the identity between the propagation of electricity and the propagation of light could be manifestly recognized in them. But the best established laws of electrostatics and electrodynamics did not lend themselves at all to the transformation dreamed up by the Scots physicist. Throughout his life, now by one route, now by another, Maxwell tried inceasingly to reduce these rebellious equations in such a fashion as to extract from them the propositions he had glimpsed, propositions which, with his marvellous genius, he guessed to be quite close to the truth. However, not one of his deductions was viable. If he finally obtained the desired equations, it was, with each new try, at the price of flagrant paralogisms or even gross faults of calculation.

There certainly was nothing of the German about this work of Maxwell's. In order to seize upon the truths his penetrating intuition revealed to him, the most impulsive and most audaciously intuitive mind that had been seen since Fresnel imposed silence on the most justified objections of the mathematical mind. The mathematical mind, in its turn, had the right and the duty to have its voice heard. Maxwell had proceeded to his discoveries by a footpath broken up by precipices impassable to every reason respectful of the rules of logic and algebra. It pertained to the mathematical mind to trace out an easy route by which it would be possible, without lacking any rigor, to rise to the same truths.

This indispensable work was carried through by a German, but by a German whose genius was exempt from the fault of the German mind. Hermann von Helmholtz showed how, without abandoning any proved truths that electrodynamics had long since mastered, without in any way offending the rules of logic and algebra, one could nevertheless attain the end the Scots physicist had proposed. All that was necessary was to impose on the propagation of the operations of electricity, not a velocity rigorously equal to that which Maxwell assigned it, but only a velocity very near to that which he assigned.

The intuitive mind and the mathematical mind were both equally satisfied in the fine theory of Helmholtz. Without disavowing any part of electrodynamics built up by Ampère, Poisson, W. Weber, and F. Neumann, his theory enriched it with everything in Maxwell's views that was true and fruitful. Helmholtz's theory, so satisfactory for all harmoniously constituted reason, was proposed by a German, and that German, celebrated for his discoveries made in the most diverse domains, enjoyed in his own land a great and legitimate renown. The theory, however, did not find any favor in Germany. Even the students of Helmholtz made nothing of it. It was one of them, Heinrich Hertz, who gave to Maxwell's thought the form in which German science from then on has taken pleasure, for the mathematical mind had rigorously driven the intuitive mind out of it.

Some objections, as numerous as they were serious, barred the way to the various methods by which Maxwell had tried to justify the equations he hoped to obtain. There was a means of cleaning away all these objections with a single stroke, a means simple to the point of brutality. This means was no longer to look upon Maxwell's equations as objects of demonstration, no more to make of them the expressions of a theory for which the commonly received laws of electrodynamics might serve as principles. It was to posit them at the outset as postulates of which algebra had only to deduce the consequences. That was what Hertz did. "The theory of Maxwell," he proclaimed, "is the very equations of Maxwell." The German mathematical mind takes singular pride in this manner of operating. In fact, in order to deduce the corollaries from equations whose origin is no longer in question, there is no need to have recourse to the intuitive mind. Algebraic calculation suffices.

That this manner of proceeding does not satisfy common sense goes without saying. The equations of Maxwell, in fact, not only run counter to the teachings of a scientific and learned physics: they directly contradict truths accessible to everyone. For whoever regards these equations as universally and rigorously true, the mere existence of a permanent magnet is inconceivable. Hertz very explicitly recognized this, as did Ludwig Boltzmann. Neither one of them, however, saw in this a sufficient reason for refusing the title of axioms to Maxwell's equations. Now, it is not just in physics laboratories that one finds permanent magnets, lodestones, needles, small bars, and horseshoes, of magnetized steel. On the bridge of every ship the binnacle contains some. One comes across them even among infants' playthings. Common sense is assuredly within its rights when it forbids the mathematical mind from denying the existence of such magnets.

Permanent magnets are also found among the instruments used by physicists who, upon Hertz's advice, receive the equations of Maxwell like orders, who submit their reason to these equations without examining their claims to such an authority. With the assistance of instruments provided by permanent magnets, these physicists execute many experiments. The results of these experiments are invoked by them when, in any specific instance, they claim to apply the corollaries of Maxwell's equations. These results tell them, then, what value it is appropriate to attribute to electrical resistance or to the coefficient of magnetization. How, then, is it possible to make use of permanent magnets at the very moment when they call on a doctrine whose axioms consider the existence of such bodies as absurd?

Such an inconsistency naturally follows the absence of the intuitive mind. Reduced to its own powers, the mathematical mind never knows how to apply its deductions to the data of experience. Between the abstractions that the theoretician considers in his processes of reasoning and the concrete bodies that the observer manipulates in the laboratory, it is the intuitive mind alone that perceives an analogy and establishes a correspondence. The connection between theoretical physics and experimental physics is intuited, not inferred.

If a theory has been composed in accordance with the laws of a sound method, if the mathematical mind and the intuitive mind have each played their legitimate roles, the relation between the equations which the mathematical mind analyzes and the facts which common sense establishes will be simple and solid. It will result from the very operations by which the intuitive mind has drawn, from the teachings of experience, the hypotheses which undergird the theory. But if the foundations of the theory have not been extracted from the entrails of reality by the intuitive mind, but are instead algebraic postulates which the mathematical mind has arbitrarily posited, there will be no natural contact any more between the consequences of theory and the results of experience. The deductions, on the one hand, and the observations, on the other, will develop in two separate domains. If some passage is established from one to the other, it will be artificially managed. The legitimacy of such transitions will be incapable of proof by virtue of the fact that the very principles of the theory have been deprived of all justification. Thus will we see the corollaries of a process of deduction applied to the objects which the axioms themselves on which this deduction depends have declared nonexistent.

IX

The study of various electrical effects has led to the supposition—indeed, it appears, to the establishment of the fact—that in the core of gases there exist very small electrically charged particles, animated with a rapid movement, which have received the name of *electrons*. By rapidly displacing in space the electrical charge which it bears, an electron acts in the fashion of an electrical current passed into a conducting body. The study of its currents is a new chapter in electrodynamics. The question is to write this chapter.

In order to compose the electrodynamics of the electron, it was possible and desirable, it seems, to follow the prudent method by which Ampère, W. Weber, and Franz Neumann had worked out the electrodynamics of the conducting body. But this method necessitated delicate experiments, penetrating intuitions, and arduous discussions, of which the works of W. Weber, Bernhard Riemann, and Clausius gave a first glimpse. It called for much ingenuity and much time. Algebraism found a means of proceeding with less trouble and more haste. The strength of previously known currents figured in the equations of Maxwell. To these were added purely and simply the strength of the *convection current* attributable to the movement of the electrons, without otherwise changing the form of the equations, and there one had the fundamental postulate of the new electro-dynamics. As soon as a Dutch physicist, M. Lorentz, had proposed this hypothesis, German scientists proceeded with an extreme ardor, to deduce from it the physics of electrons.

Thus this physics rested completely on a mere generalization of Maxwell's equations. It rested upon a beam that was known to be worm-eaten, and therefore to render decrepit the whole edifice. Bearing in themselves a formal contradiction to the mere existence of magnets, the equations of Maxwell had not been cured of this vice when the convection current was introduced. The new electrodynamics presented itself at the outset as an ensemble of corollaries of an inadmissable postulate.

This theory, vitiated by the very hypotheses which support it, did not hesitate, however, to set itself up as critic and reformer of sciences regarded up until then as the most solid. Rational mechanics, that elder sister of physical theories, which all the younger doctrines had until then taken as a guide, and from which they often even tried to draw all their principles—rational mechanics, we say, found itself shaken to its very foundations by this new arrival. In the name of electron physics, it was proposed to renounce the principles of inertia, and entirely transform the notion of mass. It was necessary if the new theory was not to be contradicted by the facts. Not for an instant did anyone ask if this contradiction, rather than necessitating the overthrow of mechanics, did not signal the incorrectness of the hypotheses upon which the electron theory rested, and did not mark the necessity of replacing or modifying them. These hypotheses were posited by the mathematical mind as postulates. It unpacked the consequences of them with an imperturbable assurance, triumphant in the very ruins which, among the doctrines established by the ages, the passage of the conquering theory piled up. Guided, nevertheless, by the experience of the past, instructed by the history of great scientific progress, the intuitive mind suspected, in this devastating march, a poor indication of truth.

Furthermore, by that inconsistency to which a reason deprived of intuition is so often condemned, the supporters of electron physics, in practice, do not scruple to make use of the very theories that their theory condemned. Their deductions require us to reject rational mechanics, but, without scruple, they had recourse to rational mechanics for the purpose of interpreting the indications of the instruments whose information they borrowed.

X

The new physics was not content to do battle with other physical theories, and in particular with rational mechanics. It did not recoil from contradicting common sense.

A delicate experiment in optics executed by M. Michelson is found to be in disaccord with electron physics, as it is, moreover, with most optical theories proposed up to our own time. In that experiment, at least if it is duly confirmed and correctly interpreted, the intuitive mind advises us to see the proof that any theory of optics developed up till now is not irreproachable, and to see the necessity of at least retouching each theory. The mathematical mind of the German physicists was of another opinion. It found a means of putting in accord the equations of the electron theory and the result of the experiment made by M. Michelson. To arrive at this conclusion, it was enough to overthrow the notions that common sense furnishes us with, concerning space and time.

The two notions of space and time appear to all men to be independent of one another. The new physics connects them to each other by an indissoluble bond. The postulate which secures this connection and which, truly, is an algebraic definition of time, has received the name of the principle of relativity. This principle of relativity, moreover, is so plainly a creation of the mathematical mind that one does not know how to articulate it correctly in ordinary language and without recourse to algebraic formulas.

One can at least show, in citing one of the consequences of the principle of relativity, at what point the connection which it establishes between the notions of space and time goes against the most explicit affirmations of common sense.

Our reason does not establish any necessary connection between the extent of the road traversed by a moving body and the time which this traversal lasts. However long the road may be, we can imagine that it could be traversed in a time as brief as we wish. However great the velocity may be, we can always conceive a greater speed. Certainly, this greater velocity may be unrealizable in fact. It may be that no physical means presently exist capable of moving a body along with a velocity greater than a given limit. But this impossibility, a limit imposed on the power of the engineer, would present no insuperable absurdity to the thought of the theoretician.

The assumptions of common sense do not hold if one admits the principle of relativity as that is conceived by Einstein, Max Abraham, Minkowski, or Laue. A body cannot move faster than light propagated in a vacuum. And this impossibility is not a simple physical impossibility, one entailed as an effect by the absence of any means capable of producing it. It is a logical impossibility. For a supporter of the principle of relativity to speak of a velocity greater than that of light is to pronounce words bereft of sense. It is to contradict the very definition of time.

That the principle of relativity disconcerts all the intuitions of common sense does not excite the distrust of German physicists. Quite the contrary, to accept it is, by that very fact, to throw over all the theories that speak of space, time, and motion, all the theories of mechanics and of physics. Such a devastation possesses nothing displeasing to German thought. On the ground which it shall have cleared of the ancient theories, the mathematical mind of the German will pleasure itself in reconstructing a complete physics, of which the principle of relativity will be the foundation. If this new physics, disdainful of common sense, goes against all that which observation and experiment have allowed us to construct in the domain of celestial and terrestrial mechanics, the purely deductive method will only be more proud of the inflexible rigor with which it will have followed, right up to the end, the ruinous consequences of its postulate.

Describing the "order of mathematics", Pascal said:

> It does not define everything and does not prove everything.
> But it only assumes things clear and established by the
> natural light, and this is why it is perfectly true since
> nature sustains it in the absence of argument. This order,
> the most perfect among men, does not at all consist in
> defining and demonstrating everything, nor in defining
> nothing and demonstrating nothing, but in holding to this
> medium of not defining things clear and understood by all
> men, and proving all other things. Against this order they
> equally sin who undertake to define and to prove everything,
> and those who neglect to do so in matters which are not
> self-evident.
>
> This is what geometry teaches perfectly. It does not define
> any of the things such as space, time, movement, number,
> equality, nor things like them, which are numerous. . . .
>
> It will be found strange perhaps that geometry is not able
> to define any of the things it has as its principal objects,
> for it cannot define either motion, number, or space. Never-
> theless, these three things are those which it particularly
> considers. . . . But there will be no surprise if it is noted
> that this admirable science only applies itself to things
> which are the most simple. This same quality, which makes
> them worthy of being its objects, makes them incapable of
> being defined, so that the lack of definition is rather a

perfection than a fault, because it does not come from
their obscurity but on the contrary from their extreme
obviousness, which is such that even if it does not have
the conviction of demonstrations, it has all of their certainty.[3]

The exclusively mathematical mind does not want to concede
to the intuitive mind the power of drawing from common
sense, where it is contained, certain knowledge endowed with
that extreme obviousness without the conviction brought by
demonstration but possessing all of its complete certitude. This
mind knows no other evidence and no other certitude than that
of definitions and demonstrations, so much so that it comes to
dream of a science where all the propositions would have been
demonstrated. And since it is contradictory to define everything
and demonstrate everything, it wishes at least to reduce all
non-defined notions and non-demonstrated judgments to the
smallest number possible. The only ideas it consents to receive
without definition are the ideas of whole number, equality,
inequality, and the addition of whole numbers. The only propositions that it will readily receive without demanding a demonstration are the axioms of arithmetic. When from such notions
and principles as these it has developed the complete doctrine
of algebra, it understands well how to reduce every science to
nothing but a chapter of that algebra. The ideas of space, of
time, of motion are presented to us by common knowledge as
simple and irreducible ideas, which cannot be reconstructed
with the aid of operations bearing on whole numbers; they are,
therefore, essentially incapable of algebraic definition. That need
be no obstacle! The mathematical mind refuses to consider the
space, time, and movement which all men conceive clearly and
about which they can discourse among themselves without ever
ceasing to understand one another. By operations referring to
algebraic expressions, that is to say, in the last analysis, on
whole numbers, it fabricates for itself its own space, its own
time, its own motion. This space, time, and motion it subjects
to postulates which are algebraic equations arbitrarily arranged.

And when, from these definitions and postulates it has, according to the rules of calculation, rigorously deduced a long series of theorems, it says that it has produced a geometry, a mechanics, a physics, although it has only developed chapters of algebra. In such a fashion was the geometry of Riemann made. So, too, was the physics of relativity formed. In such fashion does German science progress, proud of its algebraic rigidity, looking with scorn upon the good sense of which all men have received a share.

XI

Of that German science we have still only considered geometry, mechanics and physics. These are the parts of it that use mathematics incessantly; they therefore take on the algebraic form most easily. But we believe that the slightly attentive observer will once more come across the characteristics which we have recognized in the process of examining these diverse chapters of German science if he looks for them in the other chapters of this science.

No one is unaware, for example, of the extraordinary development which the study of chemistry has undergone in Germany. Now, the rise of German chemistry dates from the day on which the atomic notation was born from notions of chemical type and valence, notions given birth by the works of J. B. Dumas, Laurent, Gerhardt, Williamson, and Wurtz. This notation, in fact, with the aid of rules furnished by the part of algebra called *site analysis* [*analysis situs*], allows one to forecast, enumerate, and classify, the reactions, syntheses, and isomerisms of carbon compounds. Consequently it is the study of the compounds of carbon—organic chemistry—henceforth subject to the ascendancy of the mathematical mind, which has produced, in German laboratories, innumerable shoots of extraordinary vigor. In the numerous chapters which make up inorganic chemistry, on the contrary, the mathematical operations of the atomic notation are of quite restricted usage. The intuitive mind is still the instrument which unscrambles the complexity of reactions and classifies compounds. Consequently, these chapters of chemistry have not received from German science a tribute comparable to that paid them by French science.

We have no desire to venture into the domain of textual criticism and history. *Ne sutor ultra crepidam.* It appears, however, to our uninitiated eyes, that one could find occasion to make for these areas remarks similar to those made above.

To the taste of French science, historical studies belong essentially to the intuitive mind. The ingenuity and the lively imagination which belong to the French carried them too often, perhaps, to hazardous conclusions and to fantastic syntheses. By extolling minute research into sources and the patient verification of texts; by demanding the production of sound documents to support the least assertion, the mathematical mind of the Germans has come, quite happily, to restrain the imprudences of an excessively impulsive intuitive mind. But it has not contented itself with recalling to that mind that its power would become too fragile if it did not back up its intuitions by means of certain proofs. It has wanted to exclude the intuitive mind entirely from studies in which, up until now, it has reigned supreme. We have seen develop therefore that German erudition whose method, regulated like clockwork, claimed to lead us from texts to conclusions by infallible methods "without the least appeal to judgment and good, common sense." By the rigor of its procedures, by the systematic pace of its operations, even by the form—unintelligible to the uninitiated—of its language and the signs which it pleases to use often, this erudition clearly strove to copy the pace [*allure*] of mathematical analysis.

Now studies which require the critical sense are precisely those in which the absolute and rigid method of algebra is to the greatest extent out of place. It is above all of the examination of a historical text that one can say with Pascal:

[But in the intuitive mind] the principles are found in common
use, and are open to the scrutiny of everybody. One has
only to look, and no effort is necessary; it is only a question
of good eyesight, but the eyesight must indeed be good,
for the principles are so subtle and so numerous, that it is
almost impossible but that some escape notice. Now, the
omission of one principle leads to error; thus one must have
very clear sight to see all the principles, and then an accurate
mind not to draw false deductions from known principles.[4]

In order to retain the quite clear view of these numerous
principles which "are in common use and before the eyes of
everybody", is it reasonable to put between the eye of good
sense and the documents which one asks it to read the inextricable and closely-woven mesh-work of the German method?

XII

Must we specify the conclusion to which these various
reflections lead us? It appears to follow so naturally from what
we have just said that we feel somewhat reticent about formulating it. Therefore we shall do so with extreme brevity.

French science, German science, both deviate from ideal
and perfect science, but they deviate in two opposite ways. The
one possesses excessively that with which the other is meagerly
provided. In the one, the mathematical mind reduces the intuitive mind to the point of suffocation. In the other, the intuitive
mind dispenses too readily with the mathematical mind.

In order, therefore, that human science might develop in its
fullness and subsist in harmonious equilibrium, it would be good
to see French science and German science flourish side by side
without trying to supplant each other. Each of them ought to
understand that it finds in the other its indispensable complement.

Always, therefore, will the French find profit in pondering
the works of German scholars. They will come across either the
solid proof of truths which they have discovered and formulated

before being fully certain of them, or the refutation of the errors which an imprudent intuition had caused them to accept.

It will always be of use to Germans to study the writings of French inventors [*inventeurs*]. They will find there, so to speak, the statement of problems to the resolution of which their patient analysis ought to apply itself. They will hear there the protestations of common sense against the excess of their mathematical mind.

That German science in the nineteenth century took its departure from the work of great French thinkers no one from the other side of the Rhine would, I think, dare to contest. And no one from this side dreams of failing to recognize the contributions with which, later on, this German science has enriched our mathematics, physics, chemistry, and history.

These two sciences, then, ought to retain harmonious relations with each other. It does not follow that they are of the same rank. Intuition discovers truths; demonstration comes after and assures them. The mathematical mind gives body to the edifice the intuitive mind conceived in the first place. Between these two minds [*esprits*], there is a hierarchy analogous to that which ranks the mason in relation to the architect. The mason only does useful work if he makes it conform to the architect's plan. The mathematical mind does not pursue fruitful deductions if it does not direct them toward the end which the intuitive mind has discerned.

On the other side, for the part of the science which the deductive method constructs, the mathematical mind can well assure a rigor without reproach. But the rigor of science is not its truth. The intuitive mind alone judges if the principles of the deduction are admissible, if the consequences of demonstration are conformable to reality. For science to be true, it is not sufficient that it be rigorous; it must start from good sense, only in order to return to good sense.

The mathematical mind which inspires German science confers on it the force of a perfect discipline. But this narrowly disciplined method can only lead to disastrous results if it continues to put itself under the command of an arbitrary and senseless algebraic imperialism. The orders which it readily obeys, it ought to receive, if it wishes to do useful and beautiful work, from that which is the principal depository of good sense in the world—that is, from French science. *Scientia germanica ancilla scientiae gallicae.*

German Science and German Virtues

'Science allemande et vertus allemandes', in Gabriel Petit and Maurice Leudet (eds.),
Les allemands et la science (Paris: Librairie Felix Alcan, 1916). I have included the original
editors' introduction to Duhem's contribution.

Professor P. Duhem is one of the noblest [*plus élévé*] and most scientifically distinguished intellects of our time. The Academy of Sciences considered it an honor to make him a non-resident member, a distinction with which it is by no means lavish!

A former student of the École Normale Supérieure, M. Duhem was Lecturer in the Faculty of Sciences at Lille and then at Rennes before becoming Professor of Theoretical Physics at the University of Bordeaux (1895). His published works are numerous and highly respected: *Le potentiel thermodynamique* (1886); —*Hydrodynamique, élasticité, acoustique* (2 vols., 1891); —*Leçons sur l'électricité et le magnetisme* (3 vols., 1891–92); —*Traité élémentaire de mécanique chimique* (4 vols., 1897–99); —*Thermodynamique et chimie* (1st edn., 1902; 2nd edn., 1910); —*Les sources des théories physiques; les origines de la statique* (2 vols., 1905–06); —*La théorie physique, son objet et sa structure* (1906); —*Études sur Leonard de Vinci* (3 vols., 1906–15); —*Énergetique* (2 vols., 1911); —*Le système du monde: Histoire des doctrines cosmologiques, de Platon a Copernic* (in the course of publication; four volumes have appeared since 1913).

It should not come as a surprise that, with works as eminent as these, Professor Duhem has received honorary degrees from the Universities of Cracow and Louvain, is a member of the Dutch Society of Sciences of Haarlem, of the Royal Belgian Academy, of the Academy of Sciences of Cracow, of the Batavian Society of Experimental Sciences of Rotterdam, of the Venetian Institute of Sciences, Letters and Arts, of the Academy of Sciences, Letters and Arts of Padua, etc.

In the fine article we have before us, attention will be focussed on the striking formula by which M. Duhem as it were synthesizes his judgment: "The German imprisons the fruitful bosom of science in a corset of iron."

I

There is a German science. It is not just a collection of the work done by German scientists [*savants*]. It is further distinguished from the science of other nations by a certain number of characteristics. This is the case although these characteristics are not found in all the works which first see the light of day in Germany and are to be noticed at times in writings whose authors are not German.

Can these special signs of German science be defined with precision and be derived from certain essential dispositions of the German intellect? We think so, and on two occasions we have attempted to demonstrate this.[1] We are often reminded of the celebrated distinction which Pascal made between the intuitive mind [*l'esprit de finesse*] and the mathematical mind [*l'esprit géométrique*]. It appeared to us that one could give this description of the German intellect: In it, a serious default of the intuitive mind has allowed the mathematical mind to develop in excess.

How this default and this excess give rise to the particular forms which the sciences of reasoning, the sciences of observation, and the historical sciences have taken in Germany we have no desire to repeat here. Far from taking up again the analysis already accomplished, we might better wish to see if that analysis were not rather incomplete, if it were not conducted in too partial a fashion, if it did not overlook some of the essential characteristics of German science.

Instead of tying the particularities of that science to the qualities and defects of Teutonic reason, wouldn't it be possible just as well to tie them to the qualities and defects of the Germanic will [*volonté*]? Instead of giving an intellectual explanation, couldn't a moral explanation of the phenomenon be given?

It would be said, for example: the German is industrious; he loves work for the sake of work; he finds his enjoyment not only in the end to which the work that he does tends, but also in the work itself. Consequently, German science will not draw back from any task, no matter how arduous or extended it might be. It will excel in the accomplishment of works which can frighten away those who dread extended tiresome tasks, those who prefer to attain their goal by a short and simple route. On the contrary, the love of labor for the sake of labor, separated from the goal to which the labor tends, will often drive the German to pursue immense and toilsome researches whose purpose [*objet*] is not worth the effort necessary to attain it. He will manage to encumber the domain of science with an enormous heap of useless rubbish.

The German is meticulous. In tasks committed to him, there is no detail which he overlooks. This is why German science rises superior to all others when the most scrupulous exactitude is required. Critical editions, erudite research, everything that entails detailed observation and enumeration [*inventaire*] without missing any item—these are its predilections. On the other hand, this excessively meticulous conscience will prevent the German scholar from neglecting what is, in reality, negligible. He will rivet his myopic attention so closely on the most minute detail that he will become entirely incapable of taking in the whole entity with a single glance and perceiving its structure. "The German is capable so far as details are in question, and piteous so far as the whole is concerned," said Goethe.

The German is disciplined. It pleases him that each of his actions should be dictated by a formal and rigid rule. So, in every area of study, the German scholar will conform his manner of proceeding most exactly to the method his studies have to follow. He will not experience the impatient desire to proceed otherwise and more quickly than the rule dictated by his concern for certitude allows. On the contrary, he will appear singularly clumsy and embarrassed when the rule he follows is found to be defective, be it because that rule could not foresee the case in hand or because it foresaw something quite different. And there is no science, even geometry, so exact that the discipline does not become insufficient or dangerous when we wander from the domain which that discipline is designed to regulate. Furthermore, he will experience an excessive concern to restrict his behavior and that of others by captious and strict prescriptions. Just where everyone's initiative can without inconvenience, or even with great advantage, be allowed free play, he will want it to submit to regulation. When he shall have constrained the mathematician to make use of such-and-such notations, the chemist to employ such-and-such language, the historian to cite his reference in such-and-such a fashion, he will then be convinced that he has accomplished a useful task. He will have no repose until he has imprisoned the fruitful bosom of science in a corset of iron.

The German is submissive. To be a vassal [*homme ligé*] does not seem harsh to him. Having found favor with a master, he readily abdicates his own will. It is not only in politics that it can be said that "the fidelity of the German to his party leader is disinterested, without a preconceived or critical idea, just as the true fidelity deriving from affection ought to be."[2] The German disciple voluntarily renounces his particular advantages and his personal ambition in order to serve the advantage [*profit*] and renown of the master to whom he has given himself. Thus one sees some scientist directing and co-ordinating the work of numerous and devoted disciples much more often and much more completely in Germany than elsewhere. From their laborious, conscientious collaboration they expect nothing except the accomplishment of the work their master has conceived and the glory that ought to redound to the school and its leader. Thus are accomplished those collective works, those monuments from which the science of every land draws great profit and German science legitimate pride. On the other hand, in German schools the disciple is so accustomed to see things only through the eyes of the master that he is incapable of directly perceiving the truth any more. Though the most certain proofs have conferred upon some assertion the most manifest obviousness, he will not accept it if it contradicts the teaching he has received, or the system that has captured his faith. The saying: *Magister ipse dixit*, so severely and, at times, so unjustly brought against the universities of the Middle Ages, enjoys in the bosom of the Teutonic universities imperial sway.

Thus, it appears, the qualities and the defects of German science may be referred to the essential characteristics of the German will.

II

If we would examine the matter closely, it would become apparent that this manner of explaining the main characteristics of German science does not differ as much as might appear from that which we had set forth in the first place. Indeed, man is a unity; between his reason and his will there reigns a harmony and it sets up a mutual influence. Some qualities or defects of the heart can make the mind more or less apt for a certain manner of reasoning. Inversely, a particular cast of mind inclines the heart to a particular moral virtue or pulls it away from it.

For example, in order to use faultlessly any part of the rigid deductive method, one must not fear laborious tasks. The syllogistic enchainment by which this method proceeds is often as long as it is tedious. An attention that tires too quickly can easily cause one to lose the sequence of it. This method requires the most meticulous conscientiousness. The slightest error, the omission of the least intermediary [step], suffice to destroy completely the rigor of a demonstration, to divert the process of reasoning toward the most false of conclusions. Finally, it is impossible to practice the deductive method unless one patiently submits all the movements of one's reason to quite precise and rigid rules, whether they be of logic in general or of the particular theory of which one wishes to lay out the consequences. Is it not clear, consequently, that by his love of labor, by his meticulous attention to minute detail, by his respect for discipline, the will of the German is most favorably disposed to the development of the mathematical mind?

On the other hand, the intuitive mind gives to the searcher an advance view and a sort of foretaste of the goal toward which his effort leads. It stimulates this effort when it allows the searcher to catch sight of an object which it is worth his effort to attain. But, on the other hand, it dissuades and discourages him when it allows him to suspect that the end he pursues is not worth the effort being taken to attain it.

The intuitive mind weighs the merit of the manifold questions which present themselves to reason. Among these questions it distinguishes those that require an attentive solution. But, with many of the others, it guesses that the intellect can pass them by without needing to consider them, since they are of too slight importance to merit a response.

The intuitive mind feels that any rule, no matter how perfect it might be, does not extend to all possible circumstances. It distinguishes those cases in which it is best to escape the discipline, because to follow it would be a mistake. It is its privilege to pursue research after the truth into those regions where every rule falls away, where every prescription is silent.

The intuitive mind distinguishes between the respect due to the master and the love owed to the truth. There, where the word of the master would teach error, it can reject the received teaching. It is capable of thinking for itself and of discovering what no one taught it.

Consequently, is it not evident that a man in whom the intuitive mind is extremely well-developed will resign himself with difficulty to arduous and prolonged labor, whose object, in his eyes, holds no obvious attraction? That it will be quite difficult for him to observe a scrupulous attention to all the details, however trivial? That he will not be able to voluntarily bend the fancy of his free will to conform to the rules of a narrow discipline? That, though he be perhaps the friend of Plato, he will be even more the friend of truth, even if he has to leave the master's school? Is it not clear that the absence of the intuitive mind makes it easy for Germans to practice the moral virtues that they obviously possess?

Thus, to trace the outline of German science we could have chosen in succession two points of view which are quite different and almost opposed to one another. But between the two views [*desseins*] irreducible contrasts would not have been found. Quite the contrary, each of the two views was, so to speak, drawn from the other by a regular process and by a sort of mathematical transformation.

III

A comparison forces itself on our mind.

Hard-working to the point of loving work for its own sake and not for the result that may accompany it; meticulous to the point of not overlooking the smallest detail in the accomplishment of his task; disciplined to the point of finding his pleasure in the constraint of a rule; submissive to a superior to the point of blindly and blissfully obeying all commands issuing from his lips: such is the German by nature. But are not these natural qualities of the German will precisely the virtues that the religious order imposes on the monk?

The fear of free and self-imposed will, the need to enchain his free will to the command of a chief and the discipline of a rule explain, at one and the same time, the tendency that drives Germans to group themselves, and the characteristics which make these Teutonic groupings very similar to religious congregations.

When a German conceives the ideal organization of any association, be it an army corps, a political party, or a scientific society, he imagines it after the model of a very strictly regulated monastery. This resemblance does not escape the perception of perspicacious Germans. Prince von Bülow recalls these words of General Schweinitz: "There are only two perfect organizations on earth: the Prussian army and the Catholic Church." Bülow adds that "in that which concerns organization alone, German socialism might also deserve this eulogy."[3]

The Teuton judges rightly that he is more fit than any other man for this order founded on passive submission to the leader and on scrupulous respect for the rule. Professor Hermann

Diels writes that "the German is, here and now, on this earth, the sanctuary in which the principle of order takes refuge."[4] The German believes that this order is indispensable for every society. He dreams of imposing it, by force if necessary. This is the dream which was formulated by Professor W. Ostwald in these now-celebrated words: "Germany wants to organize Europe which, up till now, has not been organized. I shall now explain to you the great secret of Germany. We, or perhaps rather the German race, have discovered the factor of organization. Other people still live under the regimes of individualism, when we are under that of organization."

When, in a dream of the future, Professor Ostwald catches sight of the Europe he desires, Europe organized by a German triumph, he configures it entirely like one of those vast chemistry laboratories on which the universities beyond the Rhine pride themselves. There, each student punctually, scrupulously, carries out the small bit of work which the chief has entrusted to him. He does not discuss the task which he has received. He does not criticize the thought that dictated this task. He does not get tired of always doing the same measurement with the same instrument. He does not feel any desire to put some variety into his work, to exchange his habitual task for that done by some other student nearby. A toothed gear exactly meshed into a precise mechanism, he is happy to turn as the rule says he should turn, and has no concern for the finished product produced by the machine. By virtue of his natural tendencies, he lives in the laboratory to which he is attached in the same fashion as, by virtue of his vows, the Benedictine or the Carthusian lives in his monastery.

IV

If the German will behaves spontaneously in the same fashion as the will of a religious coenobite behaves by free choice; and if, on the other hand, between the behavior of the will and the behavior of the intellect a harmonious accord is established, we ought not to be astonished that German science presents numerous analogies with the learning elaborated in the bosom of monasteries.

Monks had no monopoly on Scholastic philosophy and theology. Among its masters Scholasticism counted a number of secular clerics. It is nonetheless true that, among the regular clergy, Scholasticism numbered its most illustrious and influential doctors. It is with legitimate pride that the Dominicans, the Augustinians, and the Franciscans cite the names of Thomas Aquinas, Gilles of Rome, and John Duns Scotus. In saying, then, of Scholastic thought that it was, at its best, monastic thought, one would not be mistaken. Now, was there ever a system of thought more constantly, more narrowly guided by the rules of the deductive method, a system of thought appealing less to the intuitions of the intuitive mind? Was Scholasticism not essentially, as German science is, a work of the mathematical mind? Then, again, has not the predilection with which German science has for a long time retained the vocabulary and the pace of Scholasticism been quite often noted? Do not the reasoning and the language of a Kant, for example, breathe forth a strong scent of Scholasticism which the deductions or the language of a Descartes, a Gassendi, a Malebranche no longer diffuse?

In the seventeenth century, conscientious and meticulous erudition became the attribute of the religious of the congregation of St. Maur. German criticism received from them this heritage, which it has cultivated and augmented with a veritable passion. For a hundred years, German learning has not ceased to gather up the "work of the benedictines".

Laborious, meticulous, disciplined, submissive with monastic obedience, the Germans have given to the world a science in which many essential traits of the old monastic learning can be recognized.

V

The German and the religious are both laborious, meticulous, disciplined, submissive, but they are not so in the same fashion. The German is so by virtue of instinctive tendencies which his nature imposes on his will. The religious is so by virtue of a decision by which his free will has served notice upon the rebellious parts of his nature.

From this observation a practical conclusion emerges. Every man can, if he wishes, develop the qualities which reside spontaneously in the Teutonic heart. It rests with each of us to give ourselves up to our work with a passionate ardor, to neglect nothing that would assure the perfection of that work, never through precipitateness or sluggishness to transgress the rule our work has to follow, to subdue our vanity and renounce all personal profit for the sake of the greater good of a common task. Thus is it in the power of every scientist to endow his work with the prerogatives of which German science is proud and of which it believes itself to be the exclusive possessor.

But when the man endowed with an intuitive mind gives himself over voluntarily to what the German who has not such a mind possesses naturally, the former retains a great advantage over the latter. It is open to him to acquire those dispositions which are not natural but voluntary only up to that point at which they cease to be virtues and instead become dangerous defects.

It is good that men should love the work they do, in itself, without concern for the end toward which that work leads. But this can be so only on the condition that clear-sighted thought shall have assigned some useful and worthwhile object for such extensive labor.

It is good that men should apply meticulous conscientious-ness to minute details, but on condition that a sure judgment be recognized in these minutiae, so that they are tasks worth the trouble of doing.

It is good that men follow the rule that has been traced out for them without the least infraction. But this can be so only on the condition that a logic superior to that rule shall have distinguished the circumstances under which it leads to the truth from those under which it results in error, and shall have taken care to eliminate the latter.

It is good that men submit themselves humbly to the orders of a leader, but only on condition that this leader be intelligent enough always to recognize the true and the good, and good enough so as always to desire it.

The punctual organization of a monastery is admirable only if the rule of the order has been formulated by a saint and watched over by a superior who is full of wisdom.

Now, are the natural qualities of the German apt to produce the shrewd intuition of the goal to be attained, without which all labor is so much wasted effort, or the intelligent and benevolent spontaneity of the leader, without which all discipline is abject servitude? Assuredly not. Let us grant to Professor Ostwald, if it pleases him, that Germany is capable of organizing the world after the image of a mechanism disposed with the utmost precision. He would still not be able, with assurance, either to conceive the overall task that this machine ought to accomplish or to produce the mechanic of genius who would know how to direct its movement.

The master is above the servant and the architect is above the laborer. Now, if the German has all that is requisite to be the submissive servant and the hard-working laborer, his natural qualities are not at all those of the master or the architect.

A laboratory organized like a monastery has a marvellous ability to make use of a great discovery [*découverte*], provided the director of the laboratory be an inspired inventor [*inventeur*]. But an inspired inventor never is and never can be a submissive and disciplined mind. Every invention is a revolt: a revolt against the rules which it shatters because what they prescribe is false; a revolt against the methods from which it escapes because they show themselves incapable or mendacious; a revolt against the masters, whose over-narrow teaching it extends or whose false doctrine it overthrows.

Let us apply ourselves, then, with a steadfast will to developing the love of work, conscientious concern for detail, respect for rules laid out by sound methods, submissiveness to the instructions of great masters. By such means we shall strengthen the mathematical mind in our reason, and assure to our works all the solidity, precision, rigor, and continuity on which German science prides itself.

But let us also take great care to guard and to augment the intuitive mind with which Providence has endowed us, for the intuitions of this mind will mark out for us the occasions when it is appropriate to shake the tyranny of traditions and to provide suppleness to methodical rigor. Then, smashing the bars of the cage in which it was long imprisoned, the new idea, the spontaneous and unique idea, the French idea, shall in complete liberty spread its wings and take flight.

NOTES

INTRODUCTION

1. Touched off, tellingly enough, by a remark of his daughter doubtful about the ultimate victory of France. On this and other details mentioned in this Introduction the reader will find ample documentation in my book, *Uneasy Genius: The Life and Work of Pierre Duhem* (Martinus Nijhoff: The Hague, London, and Boston, 1984).
2. As witnessed by his translation of Gilson's *From Aristotle to Darwin and Back Again: A Journey in Final Causality, Species, and Evolution* (1984) and *Linguistics and Philosophy: An Essay on the Philosophical Constants of Language* (1988), both by the University of Notre Dame Press.
3. In 1969 the University of Chicago Press published his *To Save the Phenomena*, translated by E. Dolan and C. Maschler, with an introduction by S. L. Jaki. M. Cole's translation of Duhem's *The Evolution of Mechanics*, with an introduction by B. AE. Oravas, was published by Sijthoff and Noordhoff in 1980. *Medieval Cosmology*, which is largely the translation by R. Ariew of the seventh volume of the *Système du monde*, was published by the University of Chicago Press in 1985.

LECTURE I: THE SCIENCES OF REASONING

1. Pascal, *Pensées*, art. VIII. [Trotter translation (New York: Dutton; 1958), IV, 282, p. 79. The above version is basically, however, the present translator's rendering. Throughout the text and notes, material provided by the translator has been placed in brackets. To accommodate necessary insertion, footnotes have been added when needed, and Duhem's notes thus renumbered.]
2. [René Descartes, 'Discourse on the Method of Rightly Conducting the Reason', in *The Philosophical Works of Descartes* trans. E. S. Haldane and G. R. T. Ross (Cambridge University Press; 1973), I, 81.]
3. *La Logique ou l'Art de Penser*: Part IV, ch. IX, second fault. [See Antoine Arnauld, *The Art of Thinking*, tr. James Dickoff and Patricia James (Indianapolis: Bobbs-Merrill, 1964), p. 328. This rendering is that of the present translator, however.]
4. 'Quelques reflexions sur la science allemande,' *Revue des Deux Mondes*, Feb. 1, 1915. [See the present volume.]
5. Pascal, *Pensées*, art. VIII. [Trotter translation, IV, 282, p. 79. This rendering is essentially that of the present translator, however.]

6. Pascal, *De l'esprit géométrique*, first section. [For the passage in question, see Louis Lafuma, ed., *Pascal: Oeuvres Complètes* (Paris: Editions du Seuil, 1963), p. 349.]
7. Pascal, *Pensées*, art. 1. [Trotter translation; II, 72, pp. 19–20.]
8. Pascal, *Pensées*, art. VIII [Trotter translation, VI, 395, p. 106.]
9. Kant's *Werke*, Bd. III, pp. 429–430, translated by Victor Delbos, *La philosophie pratique de Kant* (Paris: 1905), p. 232. [However, Delbos's *La philosophie pratique de Kant*, in the only edition of it available to me (Paris: Presses Universitaires de France, 1969. troisième ed., author's 'Foreword' dated 1905) is not a translation of any one of Kant's works but a commentary on most of the *Corpus*. The passage in question is not to be found on p. 232, nor anywhere near it; nor can I find it in any other likely section of the text, or in the English translation of the *Critique of Practical Reason*.]
10. [The reference is to Molière's *Les femmes savantes*. See Ronald A. Wilson and R. P. L. Ledésert, eds., *Les femmes savantes* (Boston: Heath, 1950) Acte II, Sc. VII, 21 (p. 27).]

LECTURE II: THE EXPERIMENTAL SCIENCES

1. Pascal, *Pensées*, art. VII: Difference between the mathematical mind and the intuitive mind. [Trotter translation I, 1, p. 1.]
2. Pascal, *loc. cit* [ibid., 1–2]
3. Prince von Bülow, *La politique allemande*, translated by Maurice Herbette (Paris, 1914), 'Introduction'.
4. Emmanuel Kant, *Premiers principes métaphysiques de la Science de la Nature*, translated by Ch. Andler and Ed. Chavannes (Paris: 1891); pp. 6–7. [Immanuel Kant, *Metaphysical Foundations of Natural Science*, translated by James Ellington (Indianapolis: Bobbs-Merrill; 1970), pp. 6–9.]
5. Ad. Wurtz, *Dictionnaire de Chimie*, t. 1, 1874 'Discours préliminaire'.
6. Cited by W. P. Jorrisen and L. Th. Reicher, *J.H. Van't Hoff's Amsterdamer Periode (1877–1895)* (Helder; 1912), pp. 28–29.
7. Ernest Haeckel, *Histoire de la création des être organisés d'apres les lois naturelles*, translated by Ch. Letourneau (Paris: 1874), second lesson, p. 27.
8. Spontaneous generation of living beings at the expense of purely mineral compounds.
9. Spontaneous generation at the expense of organic compounds.
10. Emphasis ours [Duhem].
11. Ernest Haeckel, *op. laud.*, Third lesson; ed. cit., p. 300.
12. In 1908, Dr. Arnold Brass sharply reproached Haeckel with having given embryological figures completely or partially invented in order to support

his theory on the simian descent of man. This accusation was the point of departure for a violent polemic.

This polemic led Haeckel to write the following lines, in which his manner of understanding and working in the natural sciences is clearly set forth:

> My enthusiasm for nature and the science of nature (which my adversaries have often characterized as fanaticism), and in particular an inclination which developed early in me to round out the entire realm of research (which several of my friends have called, in jesting, the inclination to finish things off), has often led me to go beyond the limits of exact observation and to fill in the lacunae by means of reflection and hypotheses. But I believe that I have often reached useful results precisely in this fashion, and that my philosophy of nature, which is so mocked, has done more for knowledge and for the progress of truth than the thousands of observations which I have conscientiously delivered to the public in my monographs on *radiolaria, sponges, jellyfish, siphonophores [coelenterata]*, etc. (Ernst Haeckel, *Sandalion: eine öffene Antwort auf die Falschunges-anklagen der Jesuiten.* [Frankfurt-am-Main; 1910], p. 49.)

LECTURE III: THE HISTORICAL SCIENCES

1. Fustel de Coulanges, *De la manière d'écrire l'histoire en France et en Allemagne depuis cinquante ans (Revue des Deux Mondes*, Vol. 101, 1872, p. 251).
2. [Pascal, *Pensées*, see Trotter translation, I, p. 1. The translation given, however, is the present translator's, working from Duhem's French.]
3. Aristotle, *Prior Analytics*, I, 1. U.C. 19–21 [See A. J. Jenkinson, trans., *Analytica Priora*, in Richard McKeon, ed., *The Basic Works of Aristotle* (New York: Random House 1941), p. 66: "A syllogism is discourse in which, certain things being stated, something other than what is stated follows of necessity from their being so."]
4. Fustel de Coulanges, *loc. cit.*, p. 243.
5. Fustel de Coulanges, *loc. cit.*, pp. 250–51.
6. H. Taine, *Les Origines de la France contemporaine: la Revolution*, Vol. I, *L'Anarchie*, 'Preface', p. 14.
7. Henry Houssaye, *1814*. 53rd (sic) edition (Paris: 1907), 'Preface', p. viii.
8. Fustel de Coulanges, *loc. cit.*, pp. 245, 246–47.
9. Dr. Jos. Ant. Endres, *Honorius Augustodunensis, Beitrag zur Geschichte des geistigen lebens in 12. Jahrhundert.* Kempten und München, 1906
10. Endres, *op. laud.*, p. 11.

11. Endres, *Op. laud.*, pp. 9–14.
12. [Pascal, *Pensées*, Trotter translation, IV, 282 (p. 79).]

LECTURE IV: ORDER AND CLARITY. CONCLUSION

1. *La Logique ou l'Art de penser*, IV Partie, Ch. IX, fifth fault. [Translated from Duhem's French. The translation by James Dickoff and Patricia James (*The Art of Thinking*: Indianapolis; Bobbs-Merrill; 1964, pp. 331–32) renders the passage thus:

   > *Ignoring the natural order.* Ignoring the natural order of knowledge is the geometer's greatest defect. He fancies that the only order he is called upon to observe is the order which makes the earlier propositions usable in demonstrating the later ones. The geometers treat pell-mell of lines and surface [sic] and triangles and squares, proving by complicated figures the properties of simple lines and introducing numerous inventions which mar a beautiful science. Such procedures disregard the rules of true method, which tell us always to begin with the most simple and the most general things in order to pass on to those which are more complicated and more specific.
   > Euclid often disregards the natural order.]

2. Thus we might have taken as an example a book which is excellent in other regards, the *Traité de Physique cristalline*, written by Professor Woldemar Voight. (Woldemar Voight, *Lehrbuch der Krystallphysik*. Leipzig and Berlin, 1910).

   The entire construction of this work is rigorously arranged in terms of a thought which was uttered a short time ago by Pierre Curie. This entirely geometrical thought concerned various sorts of symmetry which could affect the magnitudes intended to represent physical properties. In this work, then, two operations are joined to one another or separated from one another solely on the basis of the same kind of symmetry or symmetries of different sorts. This purely mathematical order results in placing certain phenomena which the physicist's thought constantly associates in chapters quite remote from one another. For example, the polarization and magnetizing of dielectric bodies are placed far from one another in this work. However, since Aepinus and Coulomb, the analysis of each of these two properties has not failed to reproduce the analysis of the other, and all progress in knowledge of the one has, with utter immediacy, advanced the knowledge of the other.

3. This discovery is reported in the *Mémoires de l'Académie de Berlin* for 1753.
4. Henri Poincaré, *La Valeur de la Science*, pp. 147–48. [See Henri Poincaré, *La Valeur de la Science* (Paris: Flammarion, n.d.), pp. 162–63; English

translation (George Bruce Halsted), *The Value of Science* (New York: Dover, 1958), pp. 79–80. The version in the text above is that of the present translator.]

5. Antonii Laurentii de Jussieu . . . *Genera plantarum secundum ordines naturales disposita, juxta methodum in Horto Regio Parisiensi exaratum, anno MDCCLXXIV.* Parisiis, apud Viduam Herissant et Theophilum Barrois, 1789.

6. A. L. de Jussieu, *Op. Laud.*, Introductio in Historiam plantarum, p. xxxiv.

7. A. L. de Jussieu, *loc. cit.*, p. xix.

8. A. L. de Jussieu, *loc. cit.*, p. xxxix.

9. A. L. de Jussieu, *loc. cit.*, p. xxxvii.

10. C. Saint-Saens, *Germanophilie (L'Écho de Paris*; 11 janvier, 1915).

11. Joannis Pecqueti Diepaei, *Experimenta nova anatomica, quibus incognitum hactenus chyli receptaculum, et ab eo per thoracem in ramos usque sub clavios vasa lactea deteguntur. Ejusdem Dissertatio anatomica de Circulatione sanguinis, et chyli motu.* Hardervici, apud Joannem Tollium. Juxta exemplar Parasiis impressum Anno MDCLI.

12. Pascal, *Pensées*, art. 1 [translator's rendering].

13. Lamartine, *Cours familier de Littérature*, Entretien XL, t. VII. Paris, 1859.

SOME REFLECTIONS ON GERMAN SCIENCE

1. [Blaise Pascal, *Pensées* (Trotter translation), IV, 282 (p. 79).]

2. [This text has not been available to me. I have however looked through the 1916 English translation of Émile Boutroux's *Philosophy and War* (New York: Dutton 1916). There are some interesting parallels with Duhem's thoughts here.]

3. [*De l'esprit géometrique*, in Lafuma, ed., Pascal: *Oeuvres Complètes*, pp. 350, 351.]

4. [Blaise Pascal, *Pensées*. See Trotter translation, I, 1, p. 1.]

GERMAN SCIENCE AND GERMAN VIRTUES

1. [See the previous work, 'Some Reflections on German Science', and the four lectures of *German Science*, both published in 1915.]

2. Prince de Bülow, *La Politique allemande*, (French translation), p. 148.

3. *Ibid*; 215.

4. Hermann Diels, *Internationale Monatsschrift*, 1 Nov. 1914.

INDEX

Printed in the USA
CPSIA information can be obtained
at www.ICGtesting.com
JSHW082210140824
68134JS00014B/542